ALSO BY OLIVER SACKS

Migraine

Awakenings

A Leg to Stand On

The Man Who Mistook His Wife for a Hat

Seeing Voices

An Anthropologist on Mars

The Island of the Colorblind

Uncle Tungsten

Oaxaca Journal

Musicophilia

The Mind's Eye

Hallucinations

On the Move

Gratitude

The River of Consciousness

The River of Consciousness

OLIVER SACKS

ALFRED A. KNOPF
NEW YORK · TORONTO · 2017

THIS IS A BORZOI BOOK PUBLISHED BY
ALFRED A. KNOPF AND ALFRED A. KNOPF CANADA

www.aaknopf.com
www.penguinrandomhouse.ca
www.oliversacks.com

Library of Congress Cataloging-in-Publication Data
Names: Sacks, Oliver, 1933–2015.
Title: The river of consciousness / by Oliver Sacks.
Description: New York : Alfred A. Knopf, 2017. | "A Borzoi book." | Includes bibliographical references and index.
Identifiers: LCCN 2017000815 (print) | LCCN 2017001699 (ebook) | ISBN 9780385352567 (hardcover : alk. paper) | ISBN 9780385352574 (ebook)
Subjects: LCSH: Consciousness. | Neuropsychology. | Creative ability.
Classification: LCC QP411 .S23 2017 (print) | LCC QP411 (ebook) | DDC 612.8/233—dc23
LC record available at https://lccn.loc.gov/2017000815

Library and Archives Canada Cataloguing in Publication
Sacks, Oliver, 1933–2015, author
The river of consciousness / Oliver Sacks.
Issued in print and electronic formats.
ISBN 978-0-345-80899-8
eBook ISBN 978-0-345-80901-8
1. Consciousness. 2. Neuropsychology. 3. Creative ability. I. Title.
QP411.S23 2017 612.8′233 C2017-901520-6

Jacket image: L'Eure à Pacy-sur-Eure *by Félix Vallotton, 1924. akg-images*
Jacket design by Carol Devine Carson

Manufactured in the United States of America

First Edition

For Bob Silvers

Contents

Foreword

Two weeks before his death in August 2015, Oliver Sacks outlined the contents of *The River of Consciousness,* the last book he would oversee, and charged the three of us with arranging its publication.

One of many catalysts for this book was an invitation Sacks received in 1991 from a Dutch filmmaker to participate in a documentary television series called *A Glorious Accident.* In the final episode, six scientists—the physicist Freeman Dyson, the biologist Rupert Sheldrake, the paleontologist Stephen Jay Gould, the historian of science Stephen Toulmin, the philosopher Daniel Dennett, and Dr. Sacks—gathered around a table to discuss some of the most important questions that scientists investigate: the origin of life, the meaning of evolution, the nature of consciousness. In a lively discussion, one thing was clear: Sacks could move fluidly among *all* of the disciplines. His grasp of science was not restricted to neuroscience or medicine; the issues, ideas, and questions of all the sciences enthused him. That wide-ranging expertise and passion informs the perspective of this book, in which he interrogates the nature not only of human experience but of all life (including botanical life).

In *The River of Consciousness,* he takes on evolution, botany, chemistry, medicine, neuroscience, and the arts, and calls upon his great scientific and creative heroes—above all, Darwin, Freud, and William James. For Sacks, these writers were constant companions from an early age, and much of his own work can be seen as an extended conversation with them. Like Darwin, he was an acute observer and delighted in collecting examples, many of which came from his massive correspondence with patients and colleagues. Like Freud, he was drawn to understand human behavior at its most enigmatic. And like James, even when Sacks's subject is theoretical, as in his investigations of time, memory, and creativity, his attention remains on the specificity of experience.

Dr. Sacks wished to dedicate this book to his editor, mentor, and friend of more than thirty years, Robert Silvers, who first published a number of the pieces gathered here in *The New York Review of Books.*

—Kate Edgar, Daniel Frank, and Bill Hayes

The River of Consciousness

Darwin and the Meaning of Flowers

We all know the canonical story of Charles Darwin: the twenty-two-year-old embarking on the *Beagle,* going to the ends of the earth; Darwin in Patagonia; Darwin on the Argentine pampas (managing to lasso the legs of his own horse); Darwin in South America, collecting the bones of giant extinct animals; Darwin in Australia—still a religious believer—startled at his first sight of a kangaroo ("surely two distinct Creators must have been at work"). And, of course, Darwin in the Galápagos, observing how the finches were different on each island, starting to experience the seismic shift in understanding how living things evolve that, a quarter of a century later, would result in the publication of *On the Origin of Species.*

The story climaxes here, with the publication of the *Origin* in November 1859, and has a sort of elegiac postscript: a vision of the older and ailing Darwin, in the twenty-odd years remaining to him, pottering around his gardens at Down House with no particular plan or purpose, perhaps throwing off a book or two, but with his major work long completed.

Nothing could be further from the truth. Darwin remained intensely sensitive both to criticisms and to evidence supporting his theory of natural selection, and this led him to bring out no fewer than five editions of the *Origin.* He might indeed have retreated (or returned) to his garden and his greenhouses after 1859 (there were extensive grounds around Down House, and five greenhouses), but for him these became engines of war, from which he would lob great missiles of evidence at the skeptics outside—descriptions of extraordinary structures and behaviors in plants very difficult to ascribe to special creation or design—a mass of evidence for evolution and natural selection even more overwhelming than that presented in the *Origin.*

Strangely, even Darwin scholars pay relatively little attention to this botanical work, even though it encompassed six books and seventy-odd papers. Thus Duane Isely, in his 1994 book, *One Hundred and One Botanists,* writes that while

> more has been written about Darwin than any other biologist who ever lived . . . [he] is rarely presented as a botanist. . . . The fact that he wrote several books about his research on plants is mentioned in much Darwinia, but it is casual, somewhat in the light of "Well, the great man needs to play now and then."

Darwin had always had a special, tender feeling for plants and a special admiration, too. ("It has always pleased me to exalt plants in the scale of organised beings," he wrote in his autobiography.) He grew up in a botanical family—his grandfather Erasmus Darwin had written a long, two-volume poem called *The Botanic Garden,* and Charles himself grew up in a house whose extensive gardens were filled not only with flowers but with a variety of apple trees crossbred for increased vigor. As a university student at Cambridge, the only lectures Darwin consistently attended were those of the botanist J. S. Henslow, and it was Henslow, recognizing the extraordinary qualities of his student, who recommended him for a position on the *Beagle.*

It was to Henslow that Darwin wrote very detailed letters full of observations about the fauna and flora and geology of the places he visited. (These letters, when printed and circulated, were to make Darwin famous in scientific circles even before the *Beagle* returned to England.) And it was for Henslow that Darwin, in the Galápagos, made a careful collection of all the plants in flower and noted how different islands in the archipelago could often have different species of the same genus. This was to become a crucial piece of evidence for him as he thought about the role of geographical divergence in the origin of new species.

Indeed, as David Kohn pointed out in a splendid 2008 essay, Darwin's Galápagos plant specimens, numbering well over two hundred, constituted "the single most influential natural history collection of live organisms in the entire history of science. . . . They also would turn out to be Darwin's best documented example of the evolution of species on the islands."

(The birds Darwin collected, by contrast, were not always correctly identified or labeled with their island of origin, and it was only on his return to England that these, supplemented by the specimens collected by his shipmates, were sorted out by the ornithologist John Gould.)

Darwin became close friends with two botanists, Joseph Dalton Hooker at Kew Gardens and Asa Gray at Harvard. Hooker had become his confidant in the 1840s—the only man to whom he showed the first draft of his work on evolution—and Asa Gray was to join the inner circle in the 1850s. Darwin would write to them both with increasing enthusiasm about "*our* theory."

Yet though Darwin was happy to call himself a geologist (he wrote three geological books based on his observations during the voyage of the *Beagle* and conceived a strikingly original theory on the origin of coral atolls, which was confirmed experimentally only in the second half of the twentieth century), he always insisted that he was not a botanist. One reason was that botany had (despite a precocious start in the early eighteenth century with Stephen Hales's *Vegetable Staticks,* a book full of fascinating experiments on plant physiology) remained almost entirely a descriptive and taxonomic discipline: plants were identified, classified, and named but not *investigated.* Darwin, by contrast, was preeminently an investigator, concerned with the "how" and "why" of plant structure and behavior, not just the "what."

Botany was not a mere avocation or hobby for Darwin, as it was for so many in the Victorian age; the study of plants was always infused for him with theoretical purpose, and the theoretical purpose had to do with evolution and natural selection. It was, as his son Francis wrote, "as though he were charged with theorising power ready to flow into any channel on the slightest disturbance, so that no fact, however small, could avoid releasing a stream of theory." And the flow went both ways; Darwin himself often said that "no one could be a good observer unless he was an active theoriser."

In the eighteenth century, the Swedish scientist Carl Linnaeus had shown that flowers had sexual organs (pistils and stamens), and indeed had based his classifications on these. But it was almost universally believed that flowers were self-fertilized—why else would each flower contain both male and female organs? Linnaeus himself made merry with the idea, portraying a flower with nine stamens and one pistil as a bedchamber in which a maiden was surrounded by nine lovers. A similar conceit appeared in the second volume of Darwin's grandfather's book *The Botanic Garden,* titled *The Loves of the Plants.* This was the atmosphere in which the younger Darwin grew up.

But within a year or two of his return from the *Bea-*

gle, Darwin felt forced, on theoretical grounds, to question the idea of self-fertilization. In an 1837 notebook, he wrote, "Do not plants which have male and female organs together yet receive influence from other plants?" If plants were ever to evolve, he reasoned, cross-fertilization was crucial—otherwise, no modifications could ever occur, and the world would be stuck with a single, self-reproducing plant instead of the extraordinary range of species it actually had. In the early 1840s, Darwin started to test his theory, dissecting a variety of flowers (azaleas and rhododendrons among them) and demonstrating that many of these had structural devices for preventing or minimizing self-pollination.

But it was only after *On the Origin of Species* was published in 1859 that Darwin could turn his full attention to plants. And where his early work was primarily as an observer and a collector, experiments now became his chief way of obtaining new knowledge.

He had observed, as others had, that primrose flowers came in two different forms: a "pin" form with a long style—the female part of the flower—and a "thrum" form with a short style. These differences were thought to have no particular significance. But Darwin suspected otherwise, and examining bunches of primroses that his children brought him, he found that the ratio of pins to thrums was exactly one to one.

Darwin's imagination was instantly aroused: a one-to-one ratio was what one might expect of species with separate males and females—could it be that the long-styled flowers, though hermaphrodites, were in the process of becoming female flowers and the short-styled ones male flowers? Was he actually seeing intermediate forms, evolution in action? It was a lovely idea, but it did not hold up, for the short-styled flowers, the putative males, produced as much seed as the long-styled, "female" ones. Here (as his friend T. H. Huxley would have put it) was "the slaying of a beautiful hypothesis by an ugly fact."

What, then, was the meaning of these different styles and their one-to-one ratio? Giving up theorizing, Darwin turned to experiment. Painstakingly, he tried acting as a pollinator himself, lying facedown on the lawn and transferring pollen from flower to flower: long-styled to long-styled, short-styled to short-styled, long-styled to short-styled, and vice versa. When seeds were produced, he collected and weighed them and found that the richest crop of seeds came from the crossbred flowers. He concluded that heterostyly, in which plants have styles of different length, was a special device that had evolved to facilitate outbreeding and that crossing increased the number and vitality of seeds (he called this "hybrid vigour"). Darwin later wrote, "I do not think anything in my scientific life has given me so much satisfaction as making out the meaning of the structure of these plants."

Although this subject remained a special interest of Darwin's (he published a book on it in 1877, *The Different Forms of Flowers on Plants of the Same Species*), his central concern was how flowering plants adapted themselves to using insects as agents for their own fertilization. It was well known that insects were attracted to certain flowers, visited them, and could emerge from blossoms covered with pollen. But no one had thought this was of much importance, since it was assumed that flowers were self-pollinated.

Darwin had already become suspicious of this by 1840, and in the 1850s he set five of his children to work plotting the flight routes of male humble bees. He especially admired the native orchids that grew in the meadows around Down, so he started with those. Then, with the help of friends and correspondents who sent him orchids to study, and especially Hooker, who was now director of Kew Gardens, he extended his studies to tropical orchids of all kinds.

The orchid work moved quickly and well, and in 1862 Darwin was able to send his manuscript to the printers. The book had a typically long and explicit Victorian title, *On the Various Contrivances by Which British and Foreign Orchids Are Fertilised by Insects.* His intentions, or hopes, were made clear in its opening pages:

> In my volume "On the Origin of Species" I gave only general reasons for the belief that it is an almost universal law of nature that the higher organic beings require an occasional cross with another individual. . . . I wish here to show that I have not spoken

> without having gone into details. . . . This treatise affords me also an opportunity of attempting to show that the study of organic beings may be as interesting to an observer who is fully convinced that the structure of each is due to secondary laws, as to one who views every trifling detail of structure as the result of the direct interposition of the Creator.

Here, in no uncertain terms, Darwin is throwing down the gauntlet, saying, "Explain *that* better—if you can."

Darwin interrogated orchids, interrogated flowers, as no one had ever done before, and in his orchid book he provided enormous detail, far more than is to be found in the *Origin.* This was not because he was pedantic or obsessional but because he felt that every detail was potentially significant. It is sometimes said that God is in the details, but for Darwin it was not God but natural selection, acting over millions of years, which shone out from the details, details that were unintelligible, senseless, except in the light of history and evolution. His botanical researches, his son Francis wrote,

> supplied an argument against those critics who have so freely dogmatised as to the uselessness of particular structures, and as to the consequent impossibility of their having been developed by means of natural selection. His observations on Orchids enabled him to say: "I can show the meaning of some

> of the apparently meaningless ridges and horns; who will now venture to say that this or that structure is useless?"

In a 1793 book titled *The Secret of Nature in the Form and Fertilization of Flowers Discovered,* the German botanist Christian Konrad Sprengel, a most careful observer, had noted that bees laden with pollen would carry it from one flower to another. Darwin always called this a "wonderful" book. But Sprengel, though he drew close, missed the final secret, because he was still wedded to the Linnaean idea of flowers as self-fertilizing and thought of flowers of the same species as essentially identical. It was here that Darwin made a radical break and cracked the secret of flowers, by showing that their special features—the various patterns, colors, shapes, nectars, and scents by which they lured insects to flit from one plant to another, and the devices which ensured that the insects would pick up pollen before they left the flower—were all "contrivances," as he put it; they had all evolved in the service of cross-fertilization.

What had once been a pretty picture of insects buzzing about brightly colored flowers now became an essential drama in life, full of biological depth and meaning. The colors and smells of flowers were adapted to insects' senses. While bees are attracted to blue and yellow flowers, they ignore red ones, because they are red-blind. On the other hand, their ability to see beyond the violet is exploited by flowers which use ultraviolet markings—the honey guides that direct bees to their nectaries. Butterflies, with good red vision, fertilize red flowers but may ignore the blue and violet ones. Flowers pollinated by night-flying moths tend to lack color but to exude their scents at night. And flowers pollinated by flies, which live on decaying matter, may mimic the (to us) foul smells of putrid flesh.

It was not just the evolution of plants but the *coevolution* of plants and insects that Darwin illuminated for the first time. Thus natural selection would ensure that the mouth parts of insects matched the structure of their preferred flowers—and Darwin took special delight in making predictions here. Examining one Madagascan orchid with a nectary nearly a foot long, he predicted that a moth would be found with a proboscis long enough to probe its depths; decades after his death, such a moth was finally discovered.

The *Origin* was a frontal assault (delicately presented though it was) on creationism, and while Darwin had been careful to say little in the book about human evolution, the implications of his theory were perfectly clear. It was especially the idea that man could be regarded as a mere animal—an ape—descended from other animals that had provoked outrage and ridicule. But for most people, plants were a different matter—they neither moved nor felt; they inhabited a kingdom of their own, separated from the animal kingdom by a great gulf. The evolution of plants, Darwin sensed, might seem less relevant, or less threatening, than the evolution of animals, and so more accessible to calm and rational consideration. Indeed, he wrote to Asa Gray, "no one else has perceived that my chief interest in my orchid book, has been that it was a 'flank movement' on the enemy." Darwin was never belligerent, like his "bulldog" Huxley, but he knew that there was a battle to wage, and he was not averse to military metaphors.

It is, however, not militancy or polemic that shines out of the orchid book; it is sheer joy, delight in what he was seeing. This delight and exuberance burst out of his letters:

> You cannot conceive how the Orchids have delighted me. . . . What wonderful structures! . . . The beauty of the adaptation of parts seems to me unparalleled. . . . I was almost mad at the wealth of Orchids. . . . One splendid flower of *Catasetum,* the most wonderful Orchid I have seen. . . . Happy man, he [who] has actually seen crowds of bees flying round *Catasetum,* with the pollinia sticking to their backs! . . . I never was more interested in any subject in all my life than in this of Orchids.

The fertilization of flowers engaged Darwin to the end of his life, and the orchid book was followed, nearly fifteen years later, by a more general book, *The*

Effects of Cross and Self Fertilisation in the Vegetable Kingdom.

But plants also have to survive, flourish, and find (or create) niches in the world, if they are ever to reach the point of reproduction. Darwin was equally interested in the devices and adaptations by which plants survived and in their varied and sometimes astonishing lifestyles, which included sense organs and motor powers akin to those of animals.

In 1860, during a summer holiday, Darwin first encountered and became enamored of insect-eating plants, and this started a series of investigations that culminated fifteen years later in the publication of *Insectivorous Plants.* This volume has an easy, companionable style and starts, like most of his books, with a personal recollection:

> I was surprised by finding how large a number of insects were caught by the leaves of the common sun-dew (*Drosera rotundifolia*) on a heath in Sussex. . . . On one plant all six leaves had caught their prey. . . . Many plants cause the death of insects . . . without thereby receiving, as far as we can perceive, any advantage; but it was soon evident that *Drosera* was excellently adapted for the special purpose of catching insects.

The idea of adaptation was always in Darwin's mind, and one look at the sundew showed him that these were adaptations of an entirely novel kind, for *Drosera*'s leaves not only had a sticky surface but were covered with delicate filaments (Darwin called them "tentacles") with glands at their tips. What were these for? he wondered.

"If a small organic or inorganic object be placed on the glands in the centre of a leaf," he observed,

> they transmit a motor impulse to the marginal tentacles. . . . The nearer ones are first affected and slowly bend towards the centre, and then those farther off, until at last all become closely inflected over the object.

But if the object was not nourishing, it was speedily released.

Darwin went on to demonstrate this by putting blobs of egg white on some leaves and similar blobs of inorganic matter on others. The inorganic matter was quickly released, but the egg white was retained and stimulated the formation of a ferment and an acid that soon digested and absorbed it. It was similar with insects, especially live ones. Here, without a mouth, or a gut, or nerves, *Drosera* efficiently captured its prey and absorbed it, using special digestive enzymes.

Darwin addressed not only how *Drosera* functioned but why it had adopted so extraordinary a lifestyle: he observed that the plant grew in bogs, in acidic soil that was relatively barren of organic material and assimilable nitrogen. Few plants could survive in such conditions, but *Drosera* had found a way to claim this niche by absorbing its nitrogen directly from insects rather than from the soil. Amazed by the animal-like coordination of *Drosera*'s tentacles, which closed on its prey like those of a sea anemone, and by the plant's animal-like ability to digest, Darwin wrote to Asa Gray, "You are unjust on the merits of my beloved *Drosera;* it is a wonderful plant, or rather a most sagacious animal. I will stick up for *Drosera* to the day of my death."

And he became still more enthusiastic about *Drosera* when he found that making a small nick in half of a leaf would paralyze just that half, as if a nerve had been cut. The appearance of such a leaf, he wrote, resembled "a man with his backbone broken and lower extremities paralysed." Darwin later received specimens of the Venus flytrap—a member of the sundew family—which, the moment its trigger-like hairs were touched, would clap its leaves together on an insect and imprison it. The flytrap's reactions were so fast that Darwin wondered whether electricity could be involved, something analogous to a nerve impulse. He discussed this with his physiologist colleague Burdon Sanderson and was delighted when Sanderson was able to show that electric current was indeed generated by the leaves and could also stimulate them to close. "When the leaves are irritated," Darwin recounted in *Insectivorous Plants,* "the current is disturbed in the same manner as takes place during the contraction of the muscle of an animal."

Plants are often regarded as insensate and immobile—

but the insect-eating plants provided a spectacular rebuttal of this notion, and now, eager to examine other aspects of plant motion, Darwin turned to an exploration of climbing plants. (This would culminate in the publication of *On the Movements and Habits of Climbing Plants.*) Climbing was an efficient adaptation, allowing plants to disburden themselves of rigid supporting tissue by using other plants to support and elevate them. And there was not just one way of climbing but many. There were twining plants, leaf-climbers, and plants that climbed with the use of tendrils. These especially fascinated Darwin—it was almost, he felt, as if they had "eyes" and could "survey" their surroundings for suitable supports. "I believe, Sir, the tendrils can see," he wrote to J. D. Hooker. How did such complex adaptations arise?

Darwin saw twining plants as ancestral to other climbing plants, and he thought that tendril-bearing plants had evolved from these, and leaf-climbers, in turn, from tendril-bearers, each development opening up more and more possible niches—roles for the organism in its environment. Thus climbing plants had evolved over time—they had not all been created in an instant, by divine fiat. But how did twining itself start? Darwin had observed twisting movements in the stems, leaves, and roots of every plant he had examined, and such twisting movements (which he called circumnutation) could also be observed in the earliest evolved plants: cycads, ferns, seaweeds, too. When plants grow towards the light, they do not just thrust upwards; they twist, they corkscrew, towards the light. Circumnutation, Darwin came to think, was a universal disposition of plants and the antecedent of all other twisting movements in plants.

These thoughts, along with dozens of beautiful experiments, were set out in his last botanical book, *The Power of Movement in Plants*, published in 1880. Among the many charming and ingenious experiments he recounted was one in which he planted oat seedlings, shone light on them from different directions, and found that they always bent or twisted towards the light, even when it was too dim to be seen by human eyes. Was there (as he imagined of the tips of tendrils) a photosensitive region, a sort of "eye" at the tips of the seedling leaves? He devised little caps, darkened with India ink, to cover these and found that they no longer responded to light. It was clear, he concluded, that when light fell on the leaf tip, it stimulated the tip to release some sort of messenger which, reaching the "motor" parts of the seedling, caused it to twist towards the light. Similarly, the primary roots (or radicles) of seedlings, which have to negotiate all sorts of obstacles, Darwin found to be extremely sensitive to contact, gravity, pressure, moisture, chemical gradients, etc. He wrote,

> There is no structure in plants more wonderful, as far as its functions are concerned, than the tip of the radicle. . . . It is hardly an exaggeration to say

> that the tip of the radicle . . . acts like the brain of one of the lower animals . . . receiving impressions from the sense-organs, and directing the several movements.

But as Janet Browne remarks in her biography of Darwin, *The Power of Movement in Plants* was "an unexpectedly controversial book." Darwin's idea of circumnutation was roundly criticized. He had always acknowledged it as a speculative leap, but a more cutting criticism came from the German botanist Julius Sachs, who, in Browne's words, "sneered at Darwin's suggestion that the tip of the root might be compared to the brain of a simple organism and declared that Darwin's home-based experimental techniques were laughably defective."

However homely Darwin's techniques, though, his observations were precise and correct. His ideas of a chemical messenger being transmitted downwards from the sensitive tip of the seedling to its "motor" tissue were to lead the way, fifty years later, to the discovery of plant hormones like auxins, which, in plants, play many of the roles that nervous systems do in animals.

Darwin had been an invalid for forty years, with an enigmatic illness that had assailed him since his return from the Galápagos. He would sometimes spend entire days vomiting or confined to his sofa, and as he grew older, he developed heart problems, too. But his intellectual energy and creativity never wavered. He wrote ten books after the *Origin,* many of which went through major revisions themselves—to say nothing of dozens of articles and innumerable letters. He continued to pursue his varied interests throughout his life. In 1877 he published a second edition, greatly enlarged and revised, of his orchid book (originally published fifteen years earlier). My friend Eric Korn, an antiquarian and Darwin specialist, told me that he once had a copy of this in which there was slipped the counterfoil of an 1882 postal order for two shillings and nine pence, signed by Darwin himself, in payment for a new orchid specimen. Darwin was to die in April of that year, but he was still in love with orchids and collecting them for study within weeks of his death.

Natural beauty, for Darwin, was not just aesthetic; it always reflected function and adaptation at work. Orchids were not just ornamental, to be displayed in a garden or a bouquet; they were wonderful contrivances, examples of nature's imagination, natural selection, at work. Flowers required no Creator, but were wholly intelligible as products of accident and selection, of tiny incremental changes extending over hundreds of millions of years. This, for Darwin, was the meaning of flowers, the meaning of all adaptations, plant and animal, the meaning of natural selection.

It is often felt that Darwin, more than anyone, banished "meaning" from the world—in the sense of any overall divine meaning or purpose. There is indeed no design, no plan, no blueprint in Darwin's world; natural selection has no direction or aim, nor any goal to which it strives. Darwinism, it is often said, spelled the end of teleological thinking. And yet, his son Francis writes,

> one of the greatest services rendered by my father to the study of Natural History is the revival of Teleology. The evolutionist studies the purpose or meaning of organs with the zeal of the older Teleologist, but with far wider and more coherent purpose. He has the invigorating knowledge that he is gaining not isolated conceptions of the economy of the present, but a coherent view of both past and present. And even where he fails to discover the use of any part, he may, by a knowledge of its structure, unravel the history of the past vicissitudes in the life of the species. In this way a vigour and unity is given to the study of the forms of organised beings, which before it lacked.

And this, Francis suggests, was "effected almost as much by Darwin's special botanical work as by the *Origin of Species.*"

By asking why, by seeking meaning (not in any final sense, but in the immediate sense of use or purpose), Darwin found in his botanical work the strongest evidence for evolution and natural selection. And in doing so, he transformed botany itself from a purely descriptive discipline into an evolutionary science. Botany, indeed, was the first evolutionary science, and Darwin's botanical work was to lead the way to all the other evolutionary sciences—and to the insight, as Theodosius Dobzhansky put it, that "nothing in biology makes sense except in the light of evolution."

Darwin spoke of the *Origin* as "one long argument." His botanical books, by contrast, were more personal and lyrical, less systematic in form, and they secured their effects by demonstration, not argument. According to Francis Darwin, Asa Gray observed that if the orchid book "had appeared before the *Origin*, the author would have been canonized rather than anathematized by the natural theologians."

Linus Pauling has said that he read the *Origin* before he was nine. I was not that precocious and could not have followed its "one long argument" at that age. But I had an intimation of Darwin's vision of the world in our own garden—a garden which, on summer days, was full of flowers and bees buzzing from one flower to another. It was my mother, botanically inclined, who explained to me what the bees were doing, their legs yellow with pollen, and how they and the flowers depended on each other.

While most of the flowers in the garden had rich scents and colors, we also had two magnolia trees, with huge but pale and scentless flowers. The magnolia flowers, when ripe, would be crawling with tiny insects, little beetles. Magnolias, my mother explained, were among the most ancient of flowering plants and had appeared nearly a hundred million years ago, at a time when "modern" insects like bees had not yet evolved, so they had to rely on a more ancient insect, a beetle, for pollination. Bees and butterflies, flowers with colors and scents, were not preordained, waiting in the wings—and they might never have appeared. They would develop together, in infinitesimal stages, over millions of years. The idea of a world without bees or butterflies, without scent or color, affected me with a sense of awe.

The notion of such vast eons of time—and the power of tiny, undirected changes which by their accumulation could generate new worlds, worlds of enormous richness and variety—was intoxicating. Evolutionary theory provided, for many of us, a sense of deep meaning and satisfaction that belief in a divine plan had never achieved. The world that presented itself to us became a transparent surface, through which one could see the whole history of life. The idea that it could have worked out differently, that dinosaurs might still be roaming the earth or that human beings might never have evolved, was a dizzying one. It made life seem all the more precious and a wonderful, ongoing adventure ("a glorious accident," as Stephen Jay Gould called it)—not fixed or predetermined, but always susceptible to change and new experience.

Life on our planet is several billion years old, and we literally embody this deep history in our structures, our behaviors, our instincts, our genes. We humans retain, for example, the remnants of gill arches, much modified, from our fishy ancestors and even the neural systems that once controlled gill movement. As Darwin wrote in *The Descent of Man*, "Man still bears in his bodily frame the indelible stamp of his lowly origin." We bear, too, an even older past; we are made of cells, and cells go back to the very origin of life.

In 1837, in the first of many notebooks he was to keep on "the species problem," Darwin sketched a tree of life. Its brachiating shape, so archetypal and potent, reflected the balance of evolution and extinction. Darwin always stressed the continuity of life, how all living things are descended from a common ancestor, and how we are in this sense all related to each other. So humans are related not only to apes and other animals but to plants too. (Plants and animals, we know now, share 70 percent of their DNA.) And yet, because of that great engine of natural selection—variation—every species is unique and each individual is unique, too.

The tree of life shows at a glance the antiquity and the kinship of all living organisms and how there is "descent with modification" (as Darwin originally called evolution) at every juncture. It shows too that evolution never stops, never repeats itself, never goes backwards. It shows the irrevocability of extinction—if a branch is cut off, a particular evolutionary path is lost forever.

I rejoice in the knowledge of my biological uniqueness and my biological antiquity and my biological kinship with all other forms of life. This knowledge roots me, allows me to feel at home in the natural world, to feel that I have my own sense of biological meaning, whatever my role in the cultural, human world. And although animal life is far more complex than vegetable life, and human life far more complex than the life of other animals, I trace back this sense of biological meaning to Darwin's epiphany on the meaning of flowers, and to my own intimations of this in a London garden, nearly a lifetime ago.

Speed

As a boy, I was fascinated by speed, the wild range of speeds in the world around me. People moved at different speeds; animals much more so. The wings of insects moved too fast to see, though one could judge their frequency by the tone they emitted—a hateful noise, a high E, with mosquitoes, or a lovely bass hum with the fat bumblebees that flew around the hollyhocks each summer. Our pet tortoise, which could take an entire day to cross the lawn, seemed to live in a different time frame altogether. But what then of the movement of plants? I would come down to the garden in the morning and find the hollyhocks a little higher, the roses more entwined around their trellis, but, however patient I was, I could never catch them moving.

Experiences like this played a part in turning me to photography, which allowed me to alter the rate of motion, speed it up, slow it down, so I could see, adjusted to a human perceptual rate, details of movement or change otherwise beyond the power of the eye to register. Being fond of microscopes and telescopes (my older brothers, medical students and bird-watchers, kept theirs in the house), I thought of the slowing down or the speeding up of motion as a sort of temporal equivalent: slow motion as an enlargement, a microscopy of time, and speeded-up motion as a foreshortening, a telescopy of time.

I experimented with photographing plants. Ferns, in particular, had many attractions for me, not least in their tightly wound crosiers or fiddleheads, tense with contained time, like watch springs, with the future all rolled up in them. So I would set my camera on a tripod in the garden and take photographs of fiddleheads at hourly intervals; I would develop the negatives, print them up, and bind a dozen or so prints together in a little flick-book. And then, as if by magic, I could see the fiddleheads unfurl like the curled-up paper trumpets one blew into at parties, taking a second or two for what, in real time, took a couple of days.

Slowing down motion was not so easy as speeding it up, and here I depended on my cousin, a photographer, who had a cine camera capable of taking more than a hundred frames per second. With this, I was able to catch the bumblebees at work as they hovered in the hollyhocks and to slow down their time-blurred wing beats so that I could see each up-and-down movement distinctly.

My interest in speed and movement and time, and in possible ways to make them appear faster or slower, made me take a special pleasure in two of H. G. Wells's stories, "The Time Machine" and "The New Accelerator," with their vividly imagined, almost cinematic descriptions of altered time.

"As I put on pace, night followed day like the flapping of a black wing," Wells's Time Traveller relates:

> I saw the sun hopping swiftly across the sky, leaping it every minute, and every minute marking a day. . . . The slowest snail that ever crawled dashed by too fast for me. . . . Presently, as I went on, still gaining velocity, the palpitation of night and day merged into one continuous greyness . . . the jerking sun became a streak of fire . . . the moon a fainter fluctuating band. . . . I saw trees growing and changing like puffs of vapour . . . huge buildings rise up faint and fair, and pass like dreams. The whole surface of the earth seemed changed—melting and flowing under my eyes.

The opposite of this occurs in "The New Accelerator," the story of a drug that accelerates one's perceptions, thoughts, and metabolism several thousand times or so. Its inventor and the narrator, who have taken the drug together, wander out into a glaciated world, watching

> people like ourselves and yet not like ourselves, frozen in careless attitudes, caught in mid-gesture. . . . And sliding down the air with wings flapping slowly

and at the speed of an exceptionally languid snail—
was a bee.

"The Time Machine" was published in 1895, when there was intense interest in the new powers of photography and cinematography to reveal details of movements inaccessible to the unaided eye. Étienne-Jules Marey, a French physiologist, had been the first to show that a galloping horse at one point had all four hooves off the ground. His work, as the historian Marta Braun brings out, was instrumental in stimulating Eadweard Muybridge's famous photographic studies of motion. Marey, in turn stimulated by Muybridge, went on to develop high-speed cameras that could slow and almost arrest the movements of birds and insects in flight and, at the opposite extreme, to use time-lapse photography to accelerate the otherwise almost imperceptible movements of sea urchins, starfish, and other marine animals.

I wondered sometimes whether the speeds of animals and plants could be very different from what they were: how much they were constrained by internal limits, how much by external—the gravity of the earth, the amount of energy received from the sun, the amount of oxygen in the atmosphere, and so on. So I was fascinated by yet another Wells story, *The First Men in the Moon,* with its beautiful description of how the growth of plants was dramatically accelerated on a celestial body with only a fraction of the earth's gravity:

> With a steady assurance, a swift deliberation, these amazing seeds thrust a rootlet downward to the earth and a queer little bundle-like bud into the air. . . . The bundle-like buds swelled and strained and opened with a jerk, thrusting out a coronet of little sharp tips . . . that lengthened rapidly, lengthened visibly even as we watched. The movement was slower than any animal's, swifter than any plant's I have ever seen before. How can I suggest it to you—the way that growth went on? . . . Have you ever on a cold day taken a thermometer into your warm hand and watched the little thread of mercury creep up the tube? These moon plants grew like that.

Here, as in "The Time Machine" and "The New Accelerator," the description was irresistibly cinematic and made me wonder if the young Wells had seen time-lapse photography of plants, or even experimented with it, as I had.

A few years later, when I was a student at Oxford, I read William James's *Principles of Psychology,* and there, in a wonderful chapter on "The Perception of Time," I found this description:

> We have every reason to think that creatures may possibly differ enormously in the amounts of duration which they intuitively feel, and in the fineness

of the events that may fill it. Von Baer has indulged in some interesting computations of the effect of such differences in changing the aspect of Nature. Suppose we were able, within the length of a second, to note 10,000 events distinctly, instead of barely 10, as now; if our life were then destined to hold the same number of impressions, it might be 1000 times as short. We should live less than a month, and personally know nothing of the change of seasons. If born in winter, we should believe in summer as we now believe in the heats of the Carboniferous era. The motions of organic beings would be so slow to our senses as to be inferred, not seen. The sun would stand still in the sky, the moon be almost free from change, and so on. But now reverse the hypothesis and suppose a being to get only one 1000th part of the sensations that we get in a given time, and consequently live 1000 times as long. Winters and summers will be to him like quarters of an hour. Mushrooms and the swifter-growing plants will shoot into being so rapidly as to appear instantaneous creations; annual shrubs will rise and fall from the earth like restlessly boiling water springs; the motions of animals will be as invisible as are to us the movements of bullets and cannon-balls; the sun will scour through the sky like a meteor, leaving a fiery trail behind him, etc. That such imaginary cases (barring the superhuman longevity) may be realized

somewhere in the animal kingdom, it would be rash
to deny.

This was published in 1890, when Wells was a young biologist (and writer of biology texts). Could he have read James or, for that matter, the original computations of von Baer from the 1860s? Indeed, one might say that a cinematographic model is implicit in all these descriptions, for the business of registering larger or smaller numbers of events in a given time is exactly what cine cameras do if they are run faster or slower than the usual twenty-four or so frames per second.

It is often said that time seems to go more quickly, the years rush by, as one grows older—either because when one is young one's days are packed with novel, exciting impressions or because as one grows older a year becomes a smaller and smaller fraction of one's life. But if the years appear to pass more quickly, the hours and minutes do not; they are the same as they always were.

At least they seem so to me (in my seventies), although experiments have shown that while young people are remarkably accurate at estimating a span of three minutes by counting internally, elderly subjects apparently count more slowly, so that their perceived three minutes is closer to three and a half or four minutes. But it is not clear that this phenomenon has anything to do with the existential or psychological feeling of time passing more quickly as one ages.

The hours and minutes still seem excruciatingly long when I am bored and all too short when I am engaged. As a boy, I hated school, being forced to listen passively to droning teachers. When I looked at my watch surreptitiously, counting the minutes to my liberation, the minute hand, and even the second hand, seemed to move with infinite slowness. There is an exaggerated consciousness of time in such situations; indeed, when one is bored, there may be no consciousness of anything *but* time.

In contrast were the delights of experimenting and thinking in the little chemical lab I set up at home, and here, on a weekend, I might spend an entire day in happy activity and absorption. Then I would have no consciousness of time at all, until I began to have difficulty seeing what I was doing and realized that evening had come. When, years later, I read Hannah Arendt, writing (in *The Life of the Mind*) of "a timeless region, an eternal presence in complete quiet, lying beyond human clocks and calendars altogether . . . the quiet of the Now in the time-pressed, time-tossed existence of man. . . . This small non-time space in the very heart of time," I knew exactly what she was talking about.

There have always been anecdotal accounts of people's perception of time when they are suddenly threatened with mortal danger, but the first systematic study was undertaken in 1892 by the Swiss geologist Albert Heim; he explored the mental states of thirty subjects who had survived falls in the Alps. "Mental activity became enormous, rising to a hundred-fold velocity," Heim noted. "Time became greatly expanded. . . . In many cases there followed a sudden review of the individual's entire past." In this situation, he wrote, there was "no anxiety" but rather "profound acceptance."

Almost a century later, in the 1970s, Russell Noyes and Roy Kletti, of the University of Iowa, exhumed and translated Heim's study and went on to collect and analyze more than two hundred accounts of such experiences. Most of their subjects, like Heim's, described an increased speed of thought and an apparent slowing of time during what they thought to be their last moments.

A race-car driver who was thrown thirty feet into the air in a crash said, "It seemed like the whole thing took forever. Everything was in slow motion, and it seemed to me like I was a player on a stage and could see myself tumbling over and over . . . as though I sat in the stands and saw it all happening . . . but I was not frightened." Another driver, cresting a hill at high speed and finding himself a hundred feet from a train which he was sure would kill him, observed, "As the train went by, I saw the engineer's face. It was like a movie run slowly, so that the frames progress with a jerky motion. That was how I saw his face."

While some of these near-death experiences are marked by a sense of helplessness and passivity, even dissociation, in others there is an intense sense of immediacy and reality, and a dramatic acceleration of thought and perception and reaction, which allow one to negotiate danger successfully. Noyes and Kletti describe a jet pilot who faced almost certain death when his plane was improperly launched from its carrier: "I vividly recalled, in a matter of about three seconds, over a dozen actions necessary to successful recovery of flight attitude. The procedures I needed were readily available. I had almost total recall and felt in complete control."

Many of their subjects, Noyes and Kletti said, felt that "they performed feats, both mental and physical, of which they would ordinarily have been incapable."

It may be similar, in a way, with trained athletes, especially those in games demanding fast reaction times.

A baseball may be approaching at close to a hundred miles per hour, and yet, as many people have described, the ball may seem to be almost immobile in the air, its very seams strikingly visible, and the batter finds himself in a suddenly enlarged and spacious timescape, where he has all the time he needs to hit the ball.

In a bicycle race, cyclists may be moving at nearly forty miles per hour, separated only by inches. The situation, to an onlooker, looks precarious in the extreme, and, indeed, the cyclists may be mere milliseconds away from each other. The slightest error might lead to a multiple crash. But to the cyclists themselves, concentrating intensely, everything seems to be moving in relatively slow motion, and there is ample room and time, enough to allow improvisation and intricate maneuverings.

The dazzling speed of martial arts masters, the movements too fast for the untrained eye to follow, may be executed, in the performer's mind, with an almost balletic deliberation and grace, what trainers and coaches like to call relaxed concentration. This alteration in the perception of speed is often conveyed in movies like *The Matrix* by alternating accelerated and slowed-down versions of the action.

The expertise of athletes, whatever their innate gifts, is only to be acquired by years of dedicated practice and training. At first, an intense conscious effort and attention are necessary to learn every nuance of technique and timing. But at some point the basic skills and their neural representation become so ingrained in the nervous system as to be almost second nature, no longer in need of conscious effort or decision. One level of brain activity may be working automatically, while another, the conscious level, is fashioning a perception of time, a perception which is elastic and can be compressed or expanded.

In the 1960s, the neurophysiologist Benjamin Libet, investigating how simple motor decisions were made, found that brain signals indicating an act of decision could be detected several hundred milliseconds before there was any conscious awareness of it. A champion sprinter may be up and running and already sixteen or eighteen feet into the race before he is consciously aware that the starting gun has fired. He can be off the blocks in 130 milliseconds, whereas the conscious registration of the gunshot requires 400 milliseconds or more. The runner's belief that he consciously heard the gun and then immediately exploded off the blocks is an illusion made possible, Libet would suggest, because the mind "antedates" the sound of the gun by almost half a second.

Such a reordering of time, like the apparent compression or expansion of time, raises the question of how we normally perceive time. William James speculated that our judgment of time, our speed of perception, depends on how many "events" we can perceive in a given unit of time.

There is much to suggest that conscious perception (at least visual perception) is not continuous but consists of discrete moments, like the frames of a movie, which are then blended to give an appearance of continuity. No such partitioning of time, it would seem, occurs in rapid, automatic actions such as returning a tennis shot or hitting a baseball. Christof Koch, a neuroscientist, distinguishes between "behavior" and "experience" and proposes that "behavior may be executed in a smooth fashion, while experience may be structured in discrete intervals, as in a movie." This model of consciousness would allow a Jamesian mechanism by which the perception of time could be speeded up or slowed down. Koch speculates that the apparent slowing of time in emergencies and athletic performances (at least when athletes find themselves "in the zone") may come from the power of intense attention to reduce the duration of individual frames.

For William James, the most striking departures from "normal" time were provided by the effects of certain drugs. He tried a number of them himself, from nitrous oxide to peyote, and in his chapter on the perception of time he immediately followed his meditation on von Baer with a reference to hashish. "In hashish-intoxication," he writes, "there is a curious increase in the apparent time-perspective. We utter a sentence, and ere the end is reached the beginning seems already to date from indefinitely long ago. We enter a short street, and it is as if we should never get to the end of it."

James's observations are an almost exact echo of

Jacques-Joseph Moreau's fifty years earlier. Moreau, a physician, was one of the first to make hashish fashionable in the Paris of the 1840s—indeed, he was a member, along with Gautier, Baudelaire, Balzac, and other savants and artists, of Le Club des Hachichins. Moreau wrote,

> Crossing the covered passage in the Place de l'Opéra one night, I was struck by the length of time it took to get to the other side. I had taken a few steps at most, but it seemed to me that I had been there two or three hours. . . . I hastened my step, but time did not pass more rapidly. . . . It seemed to me . . . that the walk was endlessly long and that the exit towards which I walked was retreating into the distance at the same rate as my speed of walking.

Going along with the sense that a few words, a few steps, may last an unconscionable time, there may be the sense of a world profoundly slowed, even suspended. Louis J. West, quoted in the 1970 book *Psychotomimetic Drugs* (edited by Daniel Efron), relates this anecdote: "There is a story about two hippies who are sitting in Golden Gate Park. Both are high on 'pot.' A jet aircraft goes zooming overhead and is gone; whereupon one hippie turns to the other one and says, 'Man, I thought he'd never leave!' "

But while the external world may appear slowed, an inner world of images and thoughts may take off with great speed. One may set out on an elaborate mental journey, visiting different countries and cultures, or compose a book or a symphony, or live through a whole life or an epoch of history, only to find that mere minutes or seconds have passed. Gautier described how he entered a hashish trance in which "sensations followed one another so numerous and so hurried that true appreciation of time was impossible." It seemed to him subjectively that the spell had lasted "three hundred years," but he found, on awakening, that it had lasted no more than a quarter of an hour.

The word "awakening" may be more than a figure of speech here, for such "trips" have surely to be compared with dreams or near-death experiences. I have occasionally, it seems to me, lived a whole life between my first alarm, at 5:00 a.m., and my second alarm, five minutes later.

Sometimes, as one is falling asleep, there may be a massive, involuntary jerk (a "myoclonic" jerk) of the body. Though such jerks are generated by primitive parts of the brain stem (they are, so to speak, brain-stem reflexes) and as such are without any intrinsic meaning or motive, they may be given meaning and context, turned into acts, by an instantly improvised dream. Thus the jerk may be associated with a dream of tripping or stepping over a precipice, lunging forwards to catch a ball, and so on. Such dreams may be extremely vivid and have several "scenes." Subjectively, they appear to start before the jerk, and yet presumably the entire dream mechanism is stimulated by the first preconscious perception of the jerk. All of this elaborate restructuring of time occurs in a second or less.

There are certain epileptic seizures, sometimes called experiential seizures, when a detailed recollection or hallucination of the past suddenly imposes itself upon a patient's consciousness and pursues a subjectively lengthy and unhurried course to complete itself in what, objectively, is only a few seconds. These seizures are typically associated with convulsive activity in the brain's temporal lobes and can be induced in some patients by electrical stimulation of certain trigger points on the surface of the lobes. Sometimes such epileptic experiences are suffused with a sense of metaphysical significance, along with their subjectively enormous duration. Dostoyevsky wrote of such seizures,

> There are moments, and it is only a matter of a few seconds, when you feel the presence of the eternal harmony. . . . A terrible thing is the frightful clearness with which it manifests itself and the rapture with which it fills you. . . . During these five seconds I live a whole human existence, and for that I would give my whole life and not think that I was paying too dearly.

There may be no inner sense of speed at such times, but at other times—especially with mescaline or LSD—

one may feel hurtled through thought-universes at uncontrollable, supraluminal speeds. In *The Major Ordeals of the Mind,* the French poet and painter Henri Michaux writes, "Persons returning from the speed of mescaline speak of an acceleration of a hundred or two hundred times, or even of five hundred times that of normal speed." He comments that this is probably an illusion but that even if the acceleration were much more modest—"even only six times" the normal—the increase would still feel overwhelming. What is experienced, Michaux feels, is not so much a huge accumulation of exact literal details as a series of overall impressions, dramatic highlights, as in a dream.

But, this said, if the speed of thought could be significantly heightened, the increase would readily show up (if we had the experimental means to examine it) in physiological recordings of the brain and would perhaps illustrate the limits of what is neurally possible. We would need, however, the right level of cellular activity to record from, and this would be not the level of individual nerve cells but a higher level, the level of interaction between groups of neurons in the cerebral cortex, which, in their tens or hundreds of thousands, form the neural correlate of consciousness.

The speed of such neural interactions is normally regulated by a delicate balance of excitatory and inhibitory forces, but there are certain conditions in which inhibitions may be relaxed. Dreams can take wing, move freely and swiftly, precisely because the activity of the cerebral cortex is not constrained by external perception or reality. Similar considerations, perhaps, apply to the trances induced by mescaline or hashish.

Other drugs (depressants, by and large, like opiates and barbiturates) may have the opposite effect, producing an opaque, dense inhibition of thought and movement, so that one may enter a state in which scarcely anything seems to happen and then come to, after what seems to have been a few minutes, to find that an entire day has been consumed. Such effects resemble the action of the Retarder, a drug that Wells imagined as the opposite of the Accelerator:

> The Retarder . . . should enable the patient to spread a few seconds over many hours of ordinary time, and so to maintain an apathetic inaction, a glacier-like absence of alacrity, amidst the most animated or irritating surroundings.

That there could be profound and persistent disorders of neural speed lasting for years or even decades first hit me when, in 1966, I went to work at Beth Abraham in the Bronx, a hospital for people with chronic illness, and encountered the patients whom I was later to write about in my book *Awakenings*. There were dozens of these patients in the lobby and corridors, all moving at different tempos—some violently accelerated, some in slow motion, some almost glaciated. As I looked at this landscape of disordered time, memories of Wells's Accelerator and Retarder suddenly came back to me. All of these patients, I learned, were survivors of the great pandemic of encephalitis lethargica that swept the world from 1917 to 1928. Of the millions who contracted this "sleepy sickness," about a third died in the acute stages, in states of sleep so deep as to preclude arousal, or in states of sleeplessness so intense as to preclude sedation. Some of the survivors, though often accelerated and excited in the early days, had later developed an extreme form of parkinsonism that had slowed or even frozen them, sometimes for decades. A few of the patients continued to be accelerated, and one, Ed M., was actually accelerated on one side of his body and slowed on the other.[1]

In ordinary Parkinson's disease, in addition to tremor or rigidity, one sees moderate slowings and speedings, but in postencephalitic parkinsonism, where the damage in

1. The very vocabulary of parkinsonism is couched in terms of speed. Neurologists have an array of terms to denote this: if movement is slowed, they talk about bradykinesia; if brought to a halt, akinesia; if excessively rapid, tachykinesia. Similarly, one can have bradyphrenia or tachyphrenia, a slowing or accelerating of thought.

the brain is usually far greater, there may be slowings and speedings to the utmost physiological and mechanical limits of the brain and body. Dopamine, a neurotransmitter essential for the normal flow of movement and thought, is drastically reduced in ordinary Parkinson's disease, to less than 15 percent of normal levels. In postencephalitic parkinsonism, dopamine levels may become almost undetectable.

In 1969, I was able to start most of these frozen patients on the drug L-dopa, which had recently been shown to be effective in raising dopamine levels in the brain. At first, this restored a normal speed and freedom of movement to many of the patients. But then, especially in the most severely affected, it pushed them in the opposite direction. One patient, Hester Y., showed such acceleration of movement and speech after five days on L-dopa that, I observed in my journal,

> if she had previously resembled a slow-motion film, or a persistent film frame stuck in the projector, she now gave the impression of a speeded-up film, so much so that my colleagues, looking at a film of Mrs. Y. which I took at the time, insisted that the projector was running too fast.

I assumed at first that Hester and other patients realized the unusual rates at which they were moving or speaking or thinking but were simply unable to control themselves. I soon found that this was by no means the case. Nor is it the case in patients with ordinary Parkinson's disease, as the English neurologist William Gooddy remarked at the beginning of his book *Time and the Nervous System*. An observer may note, he wrote, how slowed a parkinsonian's movements are, but "the patient will say, 'My own movements . . . seem normal unless I see how long they take by looking at a clock. The clock on the wall of the ward seems to be going exceptionally fast.'"

Gooddy refers here to "personal" time, as contrasted with "clock" time, and the extent to which personal time departs from clock time may become almost unbridgeable with the extreme bradykinesia common in postencephalitic parkinsonism. I would often see my patient Miron V. sitting in the hallway outside my office. He would appear motionless, with his right arm often lifted, sometimes an inch or two above his knee, sometimes near his face. When I questioned him about these frozen poses, he asked indignantly, "What do you mean, 'frozen poses'? I was just wiping my nose."

I wondered if he was putting me on. One morning, over a period of hours, I took a series of twenty or so photographs and stapled them together to make a flickbook, like the ones I used to make to show the unfurling of fiddleheads. With this, I could see that Miron actually was wiping his nose—but was doing so a thousand times more slowly than normal.

Hester, too, seemed unaware of the degree to which her personal time diverged from clock time. I once asked my students to play ball with her, and they found it impossible to catch her lightning-quick throws. Hester returned the ball so rapidly that their hands, still outstretched from the throw, might be hit smartly by the returning ball. "You see how quick she is," I said. "Don't underestimate her—you'd better be ready." But they could not be ready, because their best reaction times approached a seventh of a second, whereas Hester's was scarcely more than a tenth of a second.

It was only when Miron and Hester were in normal states, neither excessively retarded nor accelerated, that they could judge how startling their slowness or speed had been, and it was sometimes necessary to show them a film or a tape to convince them.[2]

With disorders of time scale, there seems almost no limit to the degree of slowing that can occur, and the speeding up of movement sometimes seems constrained only by the physical limits of articulation. If Hester tried

2. Disorders of spatial scale are as common in parkinsonism as disorders of time scale. An almost diagnostic sign of parkinsonism is micrographia—minute and often diminishingly small handwriting. Typically, patients are not aware of this at the time; it is only later, when they are back in a normal spatial frame of reference, that they are able to judge that their writing was smaller than usual. Thus there may be, for some patients, a compression of space that is comparable to the compression of time. One of my postencephalitic patients used to say, "My space, our space, is nothing like your space."

to speak or count aloud in one of her very accelerated states, the words or numbers would clash and run into each other. Such physical limitations were less evident with thought and perception. If she was shown a perspective drawing of the Necker cube (an ambiguous drawing which normally seems to switch perspective every few seconds), she might, when slowed, see switches every minute or two (or not at all, if she was "frozen"), but when speeded up, she would see the cube "flashing," changing its perspective several times a second.

Striking accelerations may also occur in Tourette's syndrome, a condition characterized by compulsions, tics, and involuntary movements and noises. Some people with Tourette's are able to catch flies on the wing. When I asked one man how he managed this, he said that he had no sense of moving especially fast but rather that, to him, the flies moved slowly.

If one reaches out a hand to touch or grasp something, the normal rate is about 1 meter per second. Normal experimental subjects, when asked to do this as quickly as possible, reach at about 4.5 meters per second. But when I asked Shane F., an artist with Tourette's, to reach as quickly as he could, he was able to achieve a rate of 7 meters per second with ease, without any sacrifice of smoothness or accuracy.[3] When I asked him to stick to

3. My colleagues and I presented these results at a meeting of the Society for Neuroscience (see Sacks, Fookson, et al., 1993).

normal speeds, his movements became constrained, awkward, inaccurate, and tic-filled.

Another patient with severe Tourette's and very rapid speech told me that in addition to the tics and vocalizations I could see and hear there were others of which—with my "slow" eyes and ears—I might be unaware. It was only with videotaping and frame-by-frame analysis that the great range of these "micro-tics" could be seen. In fact, there could be several trains of micro-tics proceeding simultaneously, apparently in complete dissociation from one another, adding up to perhaps dozens of micro-tics in a single second. The complexity of all this was as astonishing as its speed, and I thought that one could write an entire book, an atlas of tics, based on a mere five seconds of videotape. Such an atlas, I felt, would provide a sort of microscopy of the brain-mind, for all tics have determinants, whether inner or outer, and every patient's repertoire of tics is unique.

The blurted-out tics that may occur in Tourette's resemble what the great British neurologist John Hughlings Jackson called "emotional" or "ejaculate" speech (as opposed to complex, syntactically elaborate "propositional" speech). Ejaculate speech is essentially reactive, preconscious, and impulsive; it eludes the monitoring of the frontal lobes, of consciousness, and of ego, and it escapes from the mouth before it can be inhibited.

~~~~

Not just the speed but the quality of movement and thought is altered in tourettism and parkinsonism. The accelerated state tends to be exuberant in invention and fancy, leaping rapidly from one association to the next, carried along by the force of its own impetus. Slowness, in contrast, tends to go with care and caution, a sober and critical stance, which has its uses no less than the "go" of effusion. This was brought out by Ivan Vaughan, a psychologist with Parkinson's disease, who wrote a memoir called *Ivan: Living with Parkinson's Disease.* He sought to do all his writing, he told me, while he was under the influence of L-dopa, for at such times his imagination and his mental processes seemed to flow more freely and rapidly, and he had rich, unexpected associations of every sort (though if he was too accelerated, this might impair his focus and lead him to tangents in all directions). But when the effects of L-dopa wore off, he turned to editing and would find himself in a perfect state to prune the sometimes too exuberant prose he had written while he was "on."

My tourettic patient Ray, while often beleaguered and bullied by his Tourette's, also managed to exploit it in various ways. The rapidity (and sometimes oddness) of his associations made him quick-witted; he spoke of his "ticcy witticisms" and his "witty ticcicisms" and
~~~~

referred to himself as Witty Ticcy Ray.[4] This quickness and wittiness, when combined with his musical talents, made him a formidable improviser on the drums. He was almost unbeatable at Ping-Pong, partly because of his sheer speed of reaction and partly because his shots, though not technically illegal, were so unpredictable (even to himself) that his opponents were flummoxed and unable to answer them.

People with extremely severe Tourette's syndrome may be our closest approximation to the sorts of speeded-up beings imagined by von Baer and James, and people with Tourette's sometimes describe themselves as being "supercharged." "It's like having a five-hundred-horsepower engine under the hood," one of my patients says. Indeed, there are a number of world-class athletes with Tourette's—among them Jim Eisenreich and Mike Johnston in baseball, Mahmoud Abdul-Rauf in basketball, and Tim Howard in soccer.

But if the speed of Tourette's can be so adaptive, a neurological gift of sorts, then why has natural selection not served to increase the number of "speeders" in our midst? What is the sense of being relatively sluggish, staid, and "normal"? The disadvantages of excessive slowness are obvious, but it may be necessary to point out that excessive speed is equally freighted with problems. Tourettic or postencephalitic speed goes with disinhibition, an

4. Ray is described in *The Man Who Mistook His Wife for a Hat.*

impulsiveness and impetuosity that allow "inappropriate" movements and impulses to emerge precipitately. In such conditions, dangerous impulses such as putting a finger in a flame or darting in front of traffic, usually inhibited in the rest of us, may be released and acted on before consciousness can intervene.

And in extreme cases, if the stream of thought is too fast, it may lose itself, break into a torrent of superficial distractions and tangents, dissolve into a brilliant incoherence, a phantasmagoric, almost dreamlike delirium. People with severe Tourette's, like Shane, may find the movements and thoughts and reactions of other people unbearably slow, and we "neuro-normals" may at times find the Shanes of this world disconcertingly fast. "Monkeys these people seem to us," James wrote in another context, "whilst we seem to them reptilian."

In the famous chapter in *The Principles of Psychology* on "Will," James speaks of what he calls the "perverse" or pathological will and of its having two opposite forms: the "explosive" and the "obstructed." He used these terms in relation to psychological dispositions and temperaments, but they seem equally apposite in speaking of such physiological disorders as parkinsonism, Tourette's syndrome, and catatonia. (It seems strange that James never speaks of these opposites, the "explosive" and "obstructed" wills, as having, at least sometimes, a relation with each other, for he must have seen people with what we now call manic-depressive or bipolar disorder being thrown, every few weeks or months, from one extreme to the other.)

One parkinsonian friend of mine says that being in a slowed state is like being stuck in a vat of peanut butter, while being in an accelerated state is like being on ice, frictionless, slipping down an ever steeper hill, or on a tiny planet, gravityless, with no force to hold or moor him.

Though such jammed, impacted states would seem to be at the opposite pole from accelerated, explosive ones, patients can move almost instantaneously from one to the other. The term "kinesia paradoxa" was introduced by French neurologists in the 1920s to describe these remarkable if rare transitions in postencephalitic patients who had scarcely moved for years but might suddenly be "released" and move with great energy and force, only to return, after a few minutes, to their previous, motionless states. When Hester Y. was put on L-dopa, such alternations reached an extraordinary degree, and she was apt to make dozens of abrupt reversals a day.

Similar reversals may be seen in many patients with extremely severe Tourette's syndrome, who can be brought to an almost stuporous halt by the most minute dose of certain drugs. Even without medication, states of motionless and almost hypnotic concentration tend to occur in Touretters, and these represent the other side, so to speak, of the hyperactive and distractible state.

In catatonia, there may also be dramatic, instantaneous transformations from immobile, stuporous states to wildly active, frenzied ones.[5] Catatonia is rarely seen, especially in our present, tranquilized age, but some of the fear and bewilderment inspired by the insane must have come from these sudden, unpredictable transformations.

Catatonia, parkinsonism, and Tourette's, no less than manic depression, may all be thought of as "bi–polar" disorders. All of them, to use the nineteenth-century French term, are disorders *à double forme*—Janus-faced disorders which can switch incontinently from one face, one form, to the other. The possibility of any neutral state, any unpolarized state, any "normality," is so reduced in such disorders that we must envisage a dumbbell- or hourglass-shaped "surface" of disease, with only a thin neck or isthmus of neutrality between the two ends.

It is common in neurology to speak of "deficits"—the knocking out of a physiological (and perhaps psychological) function by a lesion, or area of damage, in the brain.

5. The great psychiatrist Eugen Bleuler described this in 1911:

> At times the peace and quiet is broken by the appearance of a catatonic raptus. Suddenly the patient springs up, smashes something, seizes someone with extraordinary power and dexterity. . . . A catatonic arouses himself from his rigidity, runs around the streets in his nightshirt for three hours, and finally falls down and remains lying in a cataleptic state in the gutter. The movements are often executed with great strength, and nearly always involve unnecessary muscle groups. . . . They seem to have lost control of measure and power of their movements.

Lesions in the cortex tend to produce "simple" deficits, like loss of color vision or the ability to recognize letters or numbers. In contrast, lesions in the regulatory systems of the subcortex which control movement, tempo, emotion, appetite, level of consciousness, etc., undermine control and stability, so that patients lose the normal broad base of resilience, the middle ground, and may then be thrown almost helplessly, like puppets, from one extreme to another.

~~~~

Doris Lessing once wrote of the situation of my postencephalitic patients, "It makes you aware of what a knife-edge we live on." Yet, in health, we live not on a knife edge but on a broad and stable saddleback of normality. Physiologically, neural normality reflects a balance between excitatory and inhibitory systems in the brain, a balance which, in the absence of drugs or damage, has a remarkable latitude and resilience.

We, as human beings, have relatively constant and characteristic rates of movement, though some people are a bit faster, some a bit slower, and there may be variations in our levels of energy and engagement throughout the day. We are livelier, we move a little faster, we live faster when we are young; we slow down a little, at
~~~~

least in terms of bodily movement and reaction times, as we age. But the range of all these rates, at least in ordinary people, under normal circumstances, is quite limited. There is not that much difference in reaction times between the old and the young, or between the world's best athletes and the least athletic among us. This seems to be the case with basic mental operations, too—the maximum speed at which one can perform serial computations, recognition, visual associations, and so on. The dazzling performances of chess masters, lightning-speed calculators, musical improvisers, and other virtuosos may have less to do with basic neural speed than with the vast range of knowledge, memorized patterns and strategies, and hugely sophisticated skills they can call upon.

And yet occasionally there are those who seem to reach almost superhuman speeds of thought. Robert Oppenheimer, famously, when young physicists came to explain their ideas to him, would grasp the gist and implications of their thoughts within seconds, and interrupt them, extend their thoughts, almost as soon as they opened their mouths. Virtually everyone who heard Isaiah Berlin improvise in his torrentially rapid speech, piling image upon image, idea upon idea, building enormous mental structures which evolved and dissolved before one's eyes, felt they were privy to an astonishing mental phenomenon. And this is equally so of a comic genius like Robin Williams, whose explosive, incandescent flights of association and wit seem to take off and hurtle along at rocket-like speeds. Yet here, presumably, one is dealing not with the speeds of individual nerve cells and simple circuits but with neural networks of a much higher order, exceeding the complexity of the largest supercomputer.

Nevertheless, we humans, even the fastest among us, are limited in speed by basic neural determinants, by cells with limited rates of firing, and by limited speeds of conduction between different cells and cell groups. And if somehow we could accelerate ourselves a dozen or fifty times, we would find ourselves wholly out of sync with the world around us and in a situation as bizarre as that of the narrator in Wells's story.

But we can make up for the limitations of our bodies, our senses, by using instruments of various kinds. We have unlocked time, as in the seventeenth century we unlocked space, and now have at our disposal what are, in effect, temporal microscopes and temporal telescopes of prodigious power. With these, we can achieve a quadrillion-fold acceleration or retardation, so that we can watch, by laser stroboscopy, the femtosecond-quick formation and dissolution of chemical bonds; or observe, contracted to a few minutes through computer simulation, the thirteen-billion-year history of the universe from the Big Bang to the present, or (at even higher temporal compression) its projected future to the end of time. Through such instruments, we can enhance our perceptions, speed or slow them, in effect, to a degree infinitely beyond what any living process could match. In this way, stuck though we are in our own speed and time, we can, in imagination, enter all speeds, all time.

Sentience: The Mental Lives of Plants and Worms

Charles Darwin's last book, published in 1881, was a study of the humble earthworm. His main theme—expressed in the title, *The Formation of Vegetable Mould, Through the Action of Worms*—was the immense power of worms, in vast numbers and over millions of years, to till the soil and change the face of earth.

Darwin calculated this effect:

> Nor should we forget, in considering the power which worms exert in triturating particles of rock, that there is good evidence that on each acre of land, which is sufficiently damp and not too sandy, gravelly or rocky for worms to inhabit, a weight of more than ten tons of earth annually passes through their bodies and is brought to the surface. The result for a country of the size of Great Britain, within a period not very long in a geological sense, such as a million years, cannot be insignificant.

His opening chapters, though, are devoted more simply to the "habits" of worms. Worms can distinguish between light and dark, and they generally stay underground, safe from predators, during daylight hours. They have no ears, but if they are deaf to aerial vibration, they are exceedingly sensitive to vibrations conducted through the earth, as might be generated by the footsteps of approaching animals. All of these sensations, Darwin noted, are transmitted to collections of nerve cells (he called them "the cerebral ganglia") in the worm's head.

"When a worm is suddenly illuminated," Darwin wrote, it "dashes like a rabbit into its burrow." He noted that he was "at first led to look at the action as a reflex one," but then observed that this behavior could be modified; for instance, when a worm was otherwise engaged, it showed no withdrawal with sudden exposure to light.

For Darwin, the ability to modulate responses indicated "the presence of a mind of some kind." He also wrote of the "mental qualities" of worms in relation to their plugging up their burrows, noting that "if worms are able to judge . . . having drawn an object close to the mouths of their burrows, how best to drag it in, they must acquire some notion of its general shape." This moved him to argue that worms "deserve to be called intelligent, for they then act in nearly the same manner as a man under similar circumstances."

As a boy, I played with the earthworms in our garden (and later used them in research projects), but my true love was for the seashore, and especially tidal pools, for we nearly always took our summer holidays at the seaside. This early, lyrical feeling for the beauty of simple sea creatures became more scientific under the influence of a biology teacher at school and our annual visits with him to the marine station at Millport in southwest Scotland, where we could investigate the immense range of invertebrate animals on the seashores of Cumbrae. I was so excited by these Millport visits that I thought I would like to become a marine biologist myself.

If Darwin's book on earthworms was a favorite of mine, so too was George John Romanes's 1885 book, *Jelly-Fish, Star-Fish, and Sea-Urchins: Being a Research on Primitive Nervous Systems,* with its simple, fascinating experiments and beautiful illustrations. For Romanes, Darwin's young friend and student, the seashore and its fauna were to be passionate and lifelong interests, and his aim above all was to investigate what he regarded as the behavioral manifestations of "mind" in these creatures.

I was charmed by Romanes's personal style. (His studies of invertebrate minds and nervous systems were most happily pursued, he wrote, in "a laboratory set up upon the sea-beach . . . a neat little wooden workshop thrown open to the sea-breezes.") But it was clear that correlating the neural and the behavioral was at the heart of Romanes's enterprise. He spoke of his work as "comparative psychology" and saw it as analogous to comparative anatomy.

Louis Agassiz had shown, as early as 1850, that the jellyfish *Bougainvillea* had a substantial nervous system, and by 1883 Romanes demonstrated its individual nerve cells (there are about a thousand). By simple experiments—cutting certain nerves, making incisions in the bell, or looking at isolated slices of tissue—he showed that jellyfish employed both autonomous, local mechanisms (dependent on nerve "nets") and centrally coordinated activities through the circular "brain" that ran along the margins of the bell.

By 1884, Romanes was able to include drawings of individual nerve cells and clusters of nerve cells, or ganglia, in his book *Mental Evolution in Animals.* "Throughout the animal kingdom," Romanes wrote,

> nerve tissue is invariably present in all species whose zoological position is not below that of the Hydrozoa. The lowest animals in which it has hitherto been detected are the *Medusae,* or jelly-fishes, and from them upwards its occurrence is, as I have said, invariable. Wherever it does occur its fundamental structure is very much the same, so that whether we meet with nerve-tissue in a jelly-fish, an oyster, an insect, a bird, or a man, we have no difficulty in recognizing its structural units as everywhere more or less similar.

At the same time that Romanes was vivisecting jellyfish and starfish in his seaside laboratory, the young Sigmund Freud, already a passionate Darwinian, was working in the lab of Ernst Brücke, a physiologist in Vienna. His special concern was to compare the nerve cells of vertebrates and invertebrates, in particular those of a very primitive vertebrate (*Petromyzon,* a lamprey) with those of an invertebrate (a crayfish). While it was widely held at the time that the nerve elements in invertebrate nervous systems were radically different from those of vertebrate ones, Freud was able to show and illustrate, in meticulous, beautiful drawings, that the nerve cells in crayfish were basically similar to those of lampreys—or human beings.

And he grasped, as no one had before, that the nerve cell body and its processes—dendrites and axons—constituted the basic building blocks and the signaling units of the nervous system. (Eric Kandel, in his book *In Search of Memory,* speculates that if Freud had stayed in basic research instead of going into medicine, perhaps he would be known today as "a co-founder of the neuron doctrine, instead of as the father of psychoanalysis.")

Although neurons may differ in shape and size, they are essentially the same from the most primitive animal life to the most advanced. It is their number and organization that differ: we have a hundred billion nerve cells, while a jellyfish has a thousand. But their status as cells capable of rapid and repetitive firing is essentially the same.

The crucial role of synapses—the junctions between neurons where nerve impulses can be modulated, giving organisms flexibility and a whole range of behaviors—was clarified only at the close of the nineteenth century by the great Spanish anatomist Santiago Ramón y Cajal, who looked at the nervous systems of many vertebrates and invertebrates, and by Charles Sherrington in England (it was Sherrington who coined the word "synapse" and showed that synapses could be excitatory or inhibitory in function).

In the 1880s, however, despite Agassiz's and Romanes's work, there was still a general feeling that jellyfish were little more than passively floating masses of tentacles ready to sting and ingest whatever came their way, little more than a sort of floating marine sundew.

But jellyfish are hardly passive. They pulsate rhythmically, contracting every part of their bell simultaneously, and this requires a central pacemaker system that sets off each pulse. Jellyfish can change direction and depth, and many have a "fishing" behavior that involves turning upside down for a minute, spreading their tentacles like a net, and then righting themselves, which they do by virtue of eight gravity-sensing balance organs. (If these are removed, the jellyfish is disoriented and can no longer control its position in the water.) If bitten by a fish, or otherwise threatened, jellyfish have an escape strategy—a series of rapid, powerful pulsations of the bell—that shoots them out of harm's way; special, oversize (and therefore rapidly responding) neurons are activated at such times.

Of special interest and infamous reputation among divers are the box jellyfish *(Cubomedusae)*—one of the most primitive animals to have fully developed image-forming eyes, not so different from our own. The biologist Tim Flannery wrote of box jellyfish,

> They are active hunters of medium-sized fish and crustaceans, and can move at up to twenty-one feet per minute. They are also the only jellyfish with eyes that are quite sophisticated, containing retinas, corneas, and lenses. And they have brains, which are capable of learning, memory, and guiding complex behaviors.

We and all higher animals are bilaterally symmetrical, having a front end (a head) containing a brain, and a preferred direction of movement (forwards). The jellyfish nervous system, like the animal itself, is radially symmetrical and may seem less sophisticated than a mammalian brain, but it has every right to be considered a brain, generating as it does complex adaptive behaviors and coordinating all the animal's sensory and motor mechanisms. Whether we can speak of a "mind" here (as Darwin does in regard to earthworms) depends on how one defines "mind."

~~~~
~~~~

We all distinguish between plants and animals. We understand that plants, in general, are immobile, rooted in the ground; they spread their green leaves to the heavens and feed on sunlight and soil. We understand that animals, in contrast, are mobile, moving from place to place, foraging or hunting for food; they have easily recognized behaviors of various sorts. Plants and animals have evolved along two profoundly different paths (fungi along yet another), and they are wholly different in their forms and modes of life.

And yet, Darwin insisted, they were closer than one might think. He was reinforced in this notion by the demonstration that insect-eating plants made use of electrical currents to move, just as animals did—that there was "plant electricity" as well as "animal electricity." But "plant electricity" moves slowly, roughly an inch a second, as one can see by watching the leaflets of the sensitive plant (*Mimosa pudica*) closing one by one along a leaf that is touched. "Animal electricity," conducted by nerves, moves roughly a thousand times faster.[1]

Signaling between cells depends on electrochemical changes, the flow of electrically charged atoms in and out of cells via special, highly selective molecular pores or "channels." These ion flows cause electrical currents, impulses—action potentials—that are transmitted

1. In 1852, Hermann von Helmholtz was able to measure the speed of nerve conduction at eighty feet per second. If we speed up a time-lapse film of plant movement by a thousandfold, plant behaviors start to look animal-like and may even appear "intentional."

(directly or indirectly) from one cell to another, in both plants and animals.

Plants depend largely on calcium ion channels, which suit their relatively slow lives perfectly. As Daniel Chamovitz argues in his book *What a Plant Knows*, plants are capable of registering what we would call sights, sounds, tactile signals, and much more. Plants "know" what to do, and they "remember." But without neurons, plants do not learn in the same way that animals do; instead they rely on a vast arsenal of different chemicals and what Darwin termed "devices." The blueprints for these must all be encoded in the plant's genome, and indeed plant genomes are often larger than our own.

The calcium ion channels that plants rely on do not support rapid or repetitive signaling between cells; once a plant action potential is generated, it cannot be repeated at a fast enough rate to allow, for example, the speed with which a worm "dashes . . . into its burrow." Speed requires ions and ion channels that can open and close in a matter of milliseconds, allowing hundreds of action potentials to be generated in a second. The magic ions, here, are sodium and potassium ions, which enabled the development of rapidly reacting muscle cells, nerve cells, and neuromodulation at synapses. These made possible organisms that could learn, profit by experience, judge, act, and finally think.

This new form of life—animal life—emerging perhaps 600 million years ago, conferred great advantages and transformed populations rapidly. In the so-called

Cambrian explosion (datable with remarkable precision to 542 million years ago), a dozen or more new phyla, each with very different body plans, arose within the space of a million years or less—a geological eyeblink. The once peaceful pre-Cambrian seas were transformed into a jungle of hunters and hunted, newly mobile. And while some animals (like sponges) lost their nerve cells and regressed to a vegetative life, others, especially predators, evolved increasingly sophisticated sense organs, memories, and minds.

It is fascinating to think of Darwin, Romanes, and other biologists of their time searching for "mind," "mental processes," "intelligence," even "consciousness" in primitive animals like jellyfish, and even in protozoa. A few decades afterwards, radical behaviorism would come to dominate the scene, denying reality to what was not objectively demonstrable, denying in particular any inner processes *between* stimulus and response, deeming these irrelevant or at least beyond the reach of scientific study.

Such a restriction or reduction indeed facilitated studies of stimulation and response, both with and without "conditioning," and it was Pavlov's famous studies of dogs that formalized—as "sensitization" and "habituation"—what Darwin had observed in his worms.[2]

2. Pavlov used dogs in his famous experiments on conditioned reflexes, and the conditioning stimulus was usually a bell, which the

As Konrad Lorenz wrote in *The Foundations of Ethology*, "An earthworm [that] has just avoided being eaten by a blackbird . . . is indeed well-advised to respond with a considerably lowered threshold to similar stimuli, because it is almost certain that the bird will still be nearby for the next few seconds." This lowering of threshold, or sensitization, is an elementary form of learning, even though it is nonassociative and relatively short-lived. Correspondingly, a diminution of response, or habituation, occurs when there is a repeated but insignificant stimulus—something to be ignored.

It was shown within a few years of Darwin's death that even single-cell organisms like protozoa could exhibit a range of adaptive responses. In particular, Herbert Spencer Jennings showed that the tiny, stalked, trumpet-shaped unicellular organism *Stentor* employs a repertoire of at least five different responses to being touched, before finally detaching itself to find a new site if these basic responses are ineffective. But if it is touched again, it will skip the intermediate steps and immediately take off for another site. It has become sensitized to noxious stimuli, or, to use more familiar terms, it

---

dogs learned to associate with food. But on one occasion, in 1924, there was a huge flood in the laboratory that nearly drowned the dogs. After this, many of the dogs were sensitized, even terrified, by the sight of water for the rest of their lives. Extreme or long-lasting sensitization underlies PTSD, in dogs as in humans.

"remembers" its unpleasant experience and has learned from it (though the memory lasts only a few minutes). If, conversely, *Stentor* is exposed to a series of very gentle touches, it soon ceases to respond to these at all—it has habituated.

Jennings described his work with sensitization and habituation in organisms like *Paramecium* and *Stentor* in his 1906 book *Behavior of the Lower Organisms.* Although he was careful to avoid any subjective, mentalistic language in his description of protozoan behaviors, he did include an astonishing chapter at the end of his book on the relation of observable behavior to "mind."

He felt that we humans are reluctant to attribute any qualities of mind to protozoa because they are so small:

> The writer is thoroughly convinced, after long study of the behavior of this organism, that if *Amoeba* were a large animal, so as to come within the everyday experience of human beings, its behavior would at once call forth the attribution to it of states of pleasure and pain, of hunger, desire, and the like, on precisely the same basis as we attribute these things to the dog.

Jennings's vision of a highly sensitive, dog-size *Amoeba* is almost cartoonishly the opposite of Descartes's notion of dogs as so devoid of feelings that one could vivisect them without compunction, taking their cries as purely "reflex" reactions of a quasi-mechanical kind.

Sensitization and habituation are crucial for the survival of all living organisms. These elementary forms of learning are short-lived—a few minutes at most—in protozoa and plants; longer-lived forms require a nervous system.

While behavioral studies flourished, there was almost no attention paid to the cellular basis of behavior—the exact role of nerve cells and their synapses. Investigations in mammals—involving, for example, the hippocampal or memory systems in rats—presented almost insuperable technical difficulties, due to the tiny size and extreme density of neurons (there were difficulties, moreover, even if one could record electrical activity from a single cell, in keeping it alive and fully functioning for the duration of protracted experiments).

Faced with such difficulties in his anatomical studies in the early twentieth century, Ramón y Cajal—the first and greatest microanatomist of the nervous system—had turned to study simpler systems: those of young or fetal animals, and those of invertebrates (insects, crustaceans, cephalopods, and others). For similar reasons, Eric Kandel, when he embarked in the 1960s on a study of the cellular basis of memory and learning, sought an animal with a simpler and more accessible nervous system. He settled on the giant sea snail *Aplysia*, which has 20,000 or so neurons, distributed in ten or so ganglia of about 2,000 neurons apiece. It also has particularly large neurons—some even visible to the naked eye—connected with one another in fixed anatomical circuits.

That *Aplysia* might be considered too low a form of life for studies of memory did not discountenance Kandel, despite some skepticism from his colleagues—any more than it had discountenanced Darwin when he spoke of the "mental qualities" of earthworms. "I was beginning to think like a biologist," Kandel writes, recalling his decision to work with *Aplysia.* "I appreciated that all animals have some form of mental life that reflects the architecture of their nervous system."

As Darwin had looked at an escape reflex in worms and how it might be facilitated or inhibited in different circumstances, Kandel looked at a protective reflex in *Aplysia,* the withdrawal of its exposed gill to safety, and the modulation of this response. Recording from (and sometimes stimulating) the nerve cells and synapses in the abdominal ganglion that governed these responses, he was able to show that relatively short-term memory and learning, as involved in habituation and sensitization, depended on functional changes in synapses—but longer-term memory, which might last several months, went with structural changes in the synapses. (In neither case was there any change in the actual circuits.)

As new technologies and concepts emerged in the 1970s, Kandel and his colleagues were able to complement these electrophysiological studies of memory and learning with chemical ones: "We wanted to penetrate the molecular biology of a mental process, to know exactly what molecules are responsible for short-term memory." This entailed, in particular, studies of the ion channels and neurotransmitters involved in synaptic functions—monumental work that earned Kandel a Nobel Prize.

Where *Aplysia* has only 20,000 neurons distributed in ganglia throughout its body, an insect may have up to a million nerve cells and despite its tiny size may be capable of extraordinary cognitive feats. Thus bees are expert in recognizing different colors, smells, and geometrical shapes presented in a laboratory setting, as well as systematic transformations of these. And of course, they show superb expertise in the wild or in our gardens, where they not only recognize the patterns and smells and colors of flowers but can remember their locations and communicate these to their fellow bees.

It has even been shown, in a highly social species of paper wasp, that individuals can learn and recognize the faces of other wasps. Such face learning has hitherto been described only in mammals; it is fascinating that a cognitive power so specific can be present in insects as well.

We often think of insects as tiny automata—robots with everything built in and programmed. But it is increasingly evident that insects can remember, learn, think, and communicate in quite rich and unexpected ways. Much of this, doubtless, is built in, but much, too, seems to depend on individual experience.

Whatever the case with insects, there is an altogether different situation with those geniuses among invertebrates, the cephalopods, consisting of octopuses, cuttlefish, and squid. Here, as a start, the nervous system is much larger—an octopus may have half a billion nerve cells distributed between its brain and its "arms" (a mouse, by comparison, has only 75 to 100 million). There is a remarkable degree of organization in the octopus brain, with dozens of functionally distinct lobes in the brain and similarities to the learning and memory systems of mammals.

Not only are cephalopods easily trained to discriminate test shapes and objects, but some can learn by observation, a power otherwise confined to certain birds and mammals. They have remarkable powers of camouflage and can signal complex emotions and intentions by changing their skin colors, patterns, and textures.

Darwin noted in *The Voyage of the Beagle* how an octopus in a tidal pool seemed to interact with him, by turns watchful, curious, and even playful. Octopuses can be domesticated to some extent, and their keepers often empathize with them, feeling some sense of mental and emotional proximity. Whether one can use the *c* word—"consciousness"—in regard to cephalopods can be argued all ways. But if one allows that a dog may have consciousness of a significant and individual sort, one has to allow it for an octopus, too.

Nature has employed at least two very different ways

of making a brain—indeed, there are almost as many ways as there are phyla in the animal kingdom. Mind, to varying degrees, has arisen or is embodied in all of these, despite the profound biological gulf that separates them from one other, and us from them.

The Other Road: Freud as Neurologist

It is making severe demands on the unity of the personality to try and make me identify myself with the author of the paper on the spinal ganglia of the petromyzon. Nevertheless I must be he, and I think I was happier about that discovery than about others since.

—SIGMUND FREUD TO KARL ABRAHAM,
SEPTEMBER 21, 1924

Everyone knows Freud as the father of psychoanalysis, but relatively few know about the twenty years (from 1876 to 1896) when he was primarily a neurologist and anatomist; Freud himself rarely referred to them in later life. Yet his neurological life was the precursor to his psychoanalytic one, and perhaps an essential key to it.

Freud's early and enduring passion for Darwin (along with Goethe's "Ode to Nature"), he tells us in his autobiography, made him decide to study medicine; in his first year at university, he was attending courses on "Biology and Darwinism," as well as lectures by the physiologist

Ernst Brücke. Two years later, eager to do some hands-on research, Freud asked Brücke for a position in his laboratory. Though, as Freud was later to write, he already felt that the human brain and mind might be the ultimate subject of his explorations, he was intensely curious about the early forms and origins of the nervous systems and wished to get a sense of their evolution first.

Brücke suggested that Freud look at the nervous system of a very primitive fish—*Petromyzon,* the lamprey—and in particular at the curious "Reissner" cells clustered about the spinal cord. These cells had attracted attention since Brücke's own student days forty years before, but their nature and function had never been understood. The young Freud was able to detect the precursors of these cells in the singular larval form of the lamprey, and to show they were homologous with the posterior spinal ganglia cells of higher fish—a significant discovery. (This larva of the *Petromyzon* is so different from the mature form that it was long considered to be a separate genus, *Ammocoetes.*) He then turned to studying an invertebrate nervous system, that of the crayfish. At that time the nerve "elements" of invertebrate nervous systems were considered radically different from those of vertebrate ones, but Freud was able to show that they were, in fact, morphologically identical—it was not the cellular elements that differed between primitive and advanced animals but their organization. Thus there emerged, even in Freud's earliest researches, a sense of a Darwinian evolution whereby, using the most conservative means (that is, the same basic anatomical cellular elements), more and more complex nervous systems could be built.[1]

It was natural that in the early 1880s—he now had his medical degree—Freud should move on to clinical neurology, but it was equally crucial to him that he continue his anatomical work, too, looking at human nervous systems, and he did this in the laboratory of the neuroanatomist and psychiatrist Theodor Meynert.[2] For Meynert (as for Paul Emil Flechsig and other neuroanatomists at the time) such a conjunction did not seem at all strange. There was assumed to be a simple, almost mechanical relation of mind and brain, both in health and in disease; thus Meynert's 1884 magnum opus, entitled *Psychiatry,* bore the subtitle *A Clinical Treatise on Diseases of the Fore-brain.*

Although phrenology itself had fallen into disrepute,

1. It was generally felt at this time that the nervous system was a syncytium, a continuous mass of nerve tissue, and it was not until the late 1880s and 1890s, through the efforts of Ramón y Cajal and Waldeyer, that the existence of discrete nerve cells—neurons—was appreciated. Freud, however, came very close to discovering this himself in his early studies.

2. Freud published a number of neuroanatomical studies while in Meynert's lab, focusing especially on the tracts and connections of the brain stem. He often called these anatomical studies his "real" scientific work, and he subsequently considered writing a general text on cerebral anatomy, but the book was never finished, and only a very condensed version of it was ever published, in Villaret's *Handbuch.*

the localizationist impulse had been given new life in 1861, when the French neurologist Paul Broca was able to demonstrate that a highly specific loss of function—of expressive language, a so-called expressive aphasia—followed damage to a particular part of the brain on the left side. Other correlations were quick in coming, and by the mid-1880s something akin to the phrenological dream seemed to be approaching realization, with "centers" being described for expressive language, receptive language, color perception, writing, and many other specific capabilities. Meynert reveled in this localizationist atmosphere—indeed he himself, after showing that the auditory nerves projected to a specific area of the cerebral cortex (the *Klangfeld,* or sound field), postulated that damage to this was present in all cases of sensory aphasia.

But Freud was disquieted at this theory of localization and, at a deeper level, profoundly dissatisfied, too, for he was coming to feel that all localizationism had a mechanical quality, treating the brain and nervous system as a sort of ingenious but idiotic machine, with a one-to-one correlation between elementary components and functions, denying it organization or evolution or history.

During this period (from 1882 to 1885), he spent time on the wards of the Vienna General Hospital, where he honed his skills as a clinical observer and neurologist. His heightened narrative powers, his sense of the importance of a detailed case history, are evident in the clinico-pathological papers he wrote at the time: of a boy who died from a cerebral hemorrhage associated with scurvy, an eighteen-year-old baker's apprentice with acute multiple neuritis, and a thirty-six-year-old man with a rare spinal condition—syringomyelia—who had lost the sense of pain and temperature but not the sense of touch (a dissociation caused by very circumscribed destruction within the spinal cord).

In 1886, after spending four months with the great neurologist Jean-Martin Charcot in Paris, Freud returned to Vienna to set up his own neurological practice. It is not entirely easy to reconstruct—from Freud's letters or from the vast numbers of studies and biographies of him—exactly what "neurological life" consisted of for him. He saw patients in his consulting room at 19 Berggasse, presumably a mix of patients, as might come to any neurologist then or now: some with everyday neurological disorders such as strokes, tremors, neuropathies, seizures, or migraines; and others with functional disorders such as hysterias, obsessive-compulsive conditions, or neuroses of various sorts.

He also worked at the Institute for Children's Diseases, where he held a neurological clinic several times a week. (His clinical experience here led to the books for which he became best known to his contemporaries—his three monographs on the infantile cerebral paralyses of children. These were greatly respected among the neurologists of his time and are still, on occasion, referred to even now.)

As he continued with his neurological practice, Freud's curiosity, his imagination, his theorizing powers, were on the rise, demanding more complex intellectual tasks and challenges. His earlier neurological investigations, during his years at the Vienna General Hospital, had been of a fairly conventional type, but now, as he pondered the much more complex question of the aphasias, he became convinced that a different view of the brain was needed. A more dynamic vision of the brain was taking hold of him.

It would be of great interest to know exactly how and when Freud discovered the work of the English neurologist Hughlings Jackson, who, very quietly, stubbornly, persistently, was developing an evolutionary view of the nervous system, unmoved by the localizationist frenzy all around him. Jackson, twenty years Freud's senior, had been moved to an evolutionary view of nature with the publication of Darwin's *Origin of Species* and by Herbert Spencer's evolutionary philosophy. In the early 1870s, Jackson proposed a hierarchic view of the nervous system, picturing how it might have evolved from the most primitive reflex levels, up through a series of higher and higher levels, to those of consciousness and voluntary action. In disease, Jackson conceived, this sequence was reversed, so that a dis-evolution or dissolution or regression occurred, and with this a "release" of primitive functions normally held in check by higher ones.

Although Jackson's views had first arisen in reference to certain epileptic seizures (we still speak of these as "Jacksonian" seizures), they were then applied to a variety of neurological diseases, as well as to dreams, deliriums, and insanities, and in 1879 Jackson applied them to the problem of aphasia, which had long fascinated neurologists interested in higher cognitive function.

In his own 1891 monograph *On Aphasia,* a dozen years later, Freud repeatedly acknowledged his debt to Jackson. He considered in great detail many of the special phenomena that may be seen in aphasias: the loss of new languages while the mother tongue is preserved, the preservation of the most commonly used words and associations, the retention of series of words (days of the week, for example) more than single ones, the paraphasias or verbal substitutions that might occur. Above all, he was intrigued by the stereotyped, seemingly meaningless phrases that are sometimes the sole residue of speech and which may be, as Jackson had remarked, the last utterance of the patient before his stroke. For Freud, as for Jackson, this represented the traumatic "fixation" (and thereafter the helpless repetition) of a proposition or an idea, a notion that was to assume a crucial importance in his theory of the neuroses.

Further, Freud observed that many symptoms of aphasia seemed to share associations of a more psychological rather than physiological sort. Verbal errors in aphasias might arise from verbal associations, with words of similar sound or similar meanings tending to be substituted for the correct word. But sometimes the substitution was of a more complex nature, not comprehensible as a homophone or synonym, but arising from some particular association that had been forged in the individual's past. (Here there was an intimation of Freud's later views, as set out in *The Psychopathology of Everyday Life,* of paraphasias and parapraxes as interpretable, as historically and personally meaningful.) Freud emphasized the need to look at the nature of words and their (formal or personal) associations to the universes of language and psychology, to the universe of meaning, if we wish to understand paraphasias.

He was convinced that the complex manifestations of aphasia were incompatible with any simplistic notion of word images lodged in the cells of a "center," as he wrote in *On Aphasia:*

> The theory has been evolved that the speech apparatus consists of distinct cortical centers; their cells are supposed to contain the word images (word con-

> cepts or word impressions); these centers are said to be separated by functionless cortical territory, and linked to each other by the association tracts. One may first of all raise the question as to whether such an assumption is at all correct, and even permissible. I do not believe it to be.

Instead of centers—static depots of words or images—Freud wrote, one must think of "cortical fields," large areas of cortex endowed with a variety of functions, some facilitating, some inhibiting each other. One could not make sense of the phenomena of aphasia, he continued, unless one thought in such dynamic, Jacksonian terms. Such systems, moreover, were not all at the same "level." Hughlings Jackson had suggested a vertically structured organization in the brain, with repeated representations or embodiments of function at many hierarchic levels—thus when higher-level, propositional speech has become impossible, there might still be the "regressions" characteristic of aphasia, the (sometimes explosive) emergence of primitive, emotional speech. Freud was one of the first to bring this Jacksonian notion of regression into neurology and to import it into psychiatry; one feels, indeed, that Freud's use of the concept of regression in *On Aphasia* paved the way to his much more extensive and powerful use of it in psychiatry. (One wonders what Hughlings Jackson might have thought of this vast and surprising expansion of his idea, but though he lived until 1911, we do not know whether he had ever heard of Freud.)[3]

Freud went beyond Jackson when he implied that there were no autonomous, isolable centers or functions in the brain but, rather, *systems* for achieving cognitive goals—systems that had many components and which could be created or greatly modified by the experiences of the individual. Given that literacy was not innate, for example, he felt it was not useful to think of a "center" for writing (as his friend and former colleague Sigmund Exner had postulated); one had, rather, to think of a system or systems being constructed in the brain as the result of learning (this was a striking anticipation of the notion of "functional systems" developed by A. R. Luria, the founder of neuropsychology, fifty years later).

3. If a strange silence or blindness attended Hughlings Jackson's work (his *Selected Writings* were only published in book form in 1931–32), a similar neglect attended Freud's book on aphasia. More or less ignored on publication, *On Aphasia* remained virtually unknown and unavailable for many years—even Henry Head's great monograph on aphasia, published in 1926, makes no reference to it—and was only translated into English in 1953. Freud himself spoke of *On Aphasia* as "a respectable flop" and contrasted this with the reception of his more conventional book on the cerebral paralyses of infancy:

> There is something comic about the incongruity between one's own and other people's estimation of one's work. Look at my book on the diplegias, which I knocked together almost casually, with a minimum of interest and effort. It has been a huge success. . . . But for the really good things, like the "Aphasia," the "Obsessional Ideas," which threatens to appear shortly, and the coming aetiology and theory of the neuroses, I can expect no more than a respectable flop.

In *On Aphasia,* in addition to these empirical and evolutionary considerations, Freud laid great emphasis on epistemological considerations—the confusion of categories, as he saw it, the promiscuous mixing of physical and mental:

> The relationship between the chain of physiological events in the nervous system and the mental processes is probably not one of cause and effect. The former do not cease when the latter set in . . . but, from a certain moment, a mental phenomenon corresponds to each part of the chain, or to several parts. The psychic is, therefore, a process parallel to the physiological, "a dependent concomitant."

Freud here endorsed and elaborated Jackson's views. "I do not trouble myself about the mode of connection between mind and matter," Jackson had written. "It is enough to assume a parallelism." Psychological processes have their own laws, principles, autonomies, coherences, and these must be examined independently, irrespective of whatever physiological processes may be going on in parallel. Jackson's epistemology of parallelism or concomitance gave Freud an enormous freedom to pay attention to the phenomena in unprecedented detail, to theorize, to seek a purely psychological understanding without any premature need to correlate them with physiological processes (though he never doubted that such concomitant processes must exist).

As Freud's views evolved in relation to aphasia, moving from a center or lesion way of thinking towards a dynamic view of the brain, there was an analogous movement in his views on hysteria. Charcot was convinced (and had convinced Freud at first) that although no anatomical lesions could be demonstrated in patients with *hysterical* paralyses, there must nonetheless be a "physiological lesion" (an *état dynamique*) located in the same part of the brain where, in an established *neurological* paralysis, an anatomical lesion (an *état statique*) would be found. Thus, Charcot conceived, hysterical paralyses were physiologically identical with organic ones, and hysteria could be seen, essentially, as a neurological problem, a special reactivity peculiar to certain pathologically sensitive individuals or "neuropaths."

To Freud, still saturated in anatomical and neurological thinking and very much under Charcot's spell, this seemed entirely acceptable. It was extremely difficult for him to "de-neurologize" his thinking, even in this new realm where so much was mysterious. But within a year he had become less certain. The whole neurological profession was in conflict over the question of whether hypnosis was physical or mental. In 1889, Freud paid a visit to Charcot's contemporary Hippolyte Bernheim in Nancy—Bernheim had proposed a psychological origin for hypnosis and believed that its results could be explained in terms of ideas or suggestion alone—and this seems to have influenced Freud deeply. He had begun to move away from Charcot's notion of a circumscribed (if physiological) lesion in hysterical paralysis towards a vaguer but more complex sense of physiological changes distributed among several different parts of the nervous system, a vision that paralleled the emerging insights of *On Aphasia.*

Charcot had suggested to Freud that he try to clarify the controversy by making a comparative examination of organic and hysterical paralyses.[4] This Freud was well equipped to do, for when he returned to Vienna and set up his private practice, he started to see a number of patients with hysterical paralyses and, of course, many patients with organic paralyses, too, and to attempt to elucidate their mechanisms for himself.

By 1893 he had made a complete break with all organic explanations of hysteria:

> The lesion in hysterical paralyses must be completely independent of the nervous system, since

4. The same problem was also suggested to Joseph Babinski, another young neurologist attending Charcot's clinics (and later to become one of the most famous neurologists in France). While Babinski agreed with Freud on the distinction between organic paralyses and hysterical ones, he later came to consider, when examining injured soldiers in World War I, that there was "a third realm": paralyses, anesthesias, and other neurological problems based neither on localized anatomical lesions nor on "ideas" but on broad "fields" of synaptic inhibitions in the spinal cord and elsewhere. Babinski spoke here of a "syndrome physiopathique." Such syndromes, which may follow gross physical trauma or surgical procedures, have puzzled neurologists since Silas Weir Mitchell first described them in the American Civil War, for they may incapacitate diffuse areas of the body which have neither specific innervation nor affective significance.

> in its paralyses and other manifestations hysteria behaves as though anatomy did not exist or as though it had no knowledge of it.

This was the moment of crossover, of transit, when (in a sense) Freud would give up neurology, and notions of a neurological or physiological basis for psychiatric states, and turn to looking at these states exclusively in their own terms. He was to make one final, highly theoretical attempt to delineate the neural basis of mental states in his *Project for a Scientific Psychology,* and he never gave up the notion that there must ultimately be a biological "bedrock" to all psychological conditions and theories. But for practical purposes he felt he could, and must, put these aside for a time.

~~~~

Though Freud had turned increasingly to his psychiatric work in the late 1880s and the 1890s, he continued to write occasional shorter papers on his neurological work. In 1888 he published the first description of hemianopsia in children, and in 1895 a paper on an unusual compression neuropathy (meralgia paresthetica), a condition he himself suffered from and which he had observed in several patients under his care. Freud also suffered
~~~~

from classical migraine and saw many patients with this in his neurological practice. At one point he apparently considered writing a short book on this subject, too, but, in the event, did no more than make a summary of ten "Established Points" which he sent to his friend Wilhelm Fliess in April 1895. There is a strongly physiological, quantitative tone in this summary, "an economics of nerve-force," which hinted at the extraordinary outburst of thought and writing that was to occur later in the year.

It is curious and intriguing that even with figures like Freud, who published so much, the most suggestive and prescient ideas may appear only in the course of their private letters and journals. No period in Freud's life was more productive of such ideas than the mid-1890s, when he shared the thoughts he was incubating with no one except Fliess. Late in 1895, Freud launched an ambitious attempt to bring together all his psychological observations and insights and ground them in a plausible physiology. At this point his letters to Fliess are exuberant, almost ecstatic:

> One evening last week when I was hard at work . . . the barriers were suddenly lifted, the veil drawn aside, and I had a clear vision from the details of the neuroses to the conditions that make consciousness possible. Everything seemed to connect up, the whole worked well together, and one had the impression that the Thing was now really a machine and

would soon go by itself. . . . Naturally I don't know how to contain myself for pleasure.

But this vision in which everything seemed to connect up, this vision of a complete working model of the brain and mind which presented itself to Freud with an almost revelatory lucidity, is not at all easy to grasp now (and indeed Freud himself wrote, only a few months later, "I no longer understand the state of mind in which I hatched out the 'Psychology' ").[5]

There has been intensive discussion about this *Project for a Scientific Psychology,* as it is now named (Freud's working title had been "A Psychology for Neurologists"). The *Project* makes very difficult reading, partly because of the intrinsic difficulty and originality of many of its concepts; partly because Freud uses outmoded and sometimes idiosyncratic terms that we have to translate into more familiar ones; partly because it was written at furious speed in a sort of shorthand; and perhaps because it might never have been intended for anyone's eyes but his own.

And yet the *Project* does bring together, or attempts to bring together, the domains of memory, attention, consciousness, perception, wishes, dreams, sexuality,

5. Freud never reclaimed his manuscript from Fliess, and it was presumed lost until the 1950s, when it was finally found and published—although what was found was only a fragment of the many drafts Freud wrote in late 1895.

defense, repression, and primary and secondary thought processes (as he called them) into a single coherent vision of the mind, and to ground all of these processes in a basic physiological framework, constituted by different systems of neurons, their interactions and modifiable "contact barriers," and free and bound states of neural excitation.

Though the language of the *Project* is inevitably that of the 1890s, a number of its notions retain (or have assumed) striking relevance to many current ideas in neuroscience, and this has caused it to be reexamined by Karl Pribram and Merton Gill, among others. Pribram and Gill, indeed, call the *Project* "a Rosetta stone" for those who wish to make connections between neurology and psychology. Many of the ideas Freud advanced in the *Project,* moreover, can now be examined experimentally in ways that would have been impossible at the time they were formulated.

The nature of memory occupied Freud from first to last. Aphasia he saw as a sort of forgetting, and he had observed in his notes that an early symptom in migraine was often the forgetting of proper names. He saw a pathology of memory as central in hysteria ("Hysterics suffer mainly from reminiscences"), and in the *Project* he attempted to explicate the physiological basis of memory at many levels. One physiological prerequisite for memory, he postulated, was a system of "contact barriers" between certain neurons—his so-called psi system (this was a decade before Sherrington gave synapses their name). Freud's contact barriers were capable of selective facilitation or inhibition, thus allowing permanent neuronal changes that corresponded to the acquisition of new information and new memories—a theory of learning basically similar to one that Donald Hebb would propose in the 1940s and which is now supported by experimental findings.

At a higher level, Freud regarded memory and motive as inseparable. Recollection could have no force, no meaning, unless it was allied with motive. The two had always to be coupled together, and in the *Project,* as Pribram and Gill emphasize, "both memory and motive are psi processes based on selective facilitation . . . memories [being] the retrospective aspect of these facilitations; motives the prospective aspects."[6]

Thus remembering, for Freud, though it required such local neuronal traces (of the sort we now call long-term

6. The inseparability of memory and motive, Freud pointed out, opened the possibility of understanding certain *illusions* of memory based on intentionality: the illusion that one has written to a person, for instance, when one has not but intended to, or that one has run the bath when one has merely intended to do so. We do not have such illusions unless there has been a preceding intention.

potentiation), went far beyond them and was essentially a dynamic, transforming, reorganizing process throughout the course of life. Nothing was more central to the formation of identity than the power of memory; nothing more guaranteed one's continuity as an individual. But memories shift, and no one was more sensitive than Freud to the reconstructive potential of memory, to the fact that memories are continually worked over and revised and that their essence, indeed, *is* recategorization.

Arnold Modell has taken up this point with regard to both the therapeutic potential of psychoanalysis and, more generally, the formation of a private self. He quotes a letter Freud wrote to Fliess in December 1896 in which he used the term *Nachträglichkeit,* which Modell feels is most accurately rendered as "retranscription."

"As you know," Freud wrote,

> I am working on the assumption that our psychic mechanism has come into being by a process of stratification, the material present in the form of memory traces being subjected from time to time to a *rearrangement* in accordance with fresh circumstances—a *retranscription.* . . . Memory is present not once but several times over . . . the successive registrations representing the psychic achievement of successive epochs of life. . . . I explain the peculiarities of the psychoneuroses by supposing that this translation has not taken place in the case of some of the material.

The potential for therapy, for change, therefore, lies in the capacity to exhume such "fixated" material into the present so that it can be subjected to the creative process of retranscription, allowing the stalled individual to grow and change once again.

Such remodelings, Modell feels, not only are crucial in the therapeutic process but are a constant part of human life both for day-to-day "updating" (an updating which those with amnesia cannot do) and for the major (and sometimes cataclysmic) transformations, the "revaluations of all values" (as Nietzsche would say) which are necessary for the evolution of a unique private self.

That memory does construct and reconstruct, endlessly, was a central conclusion of the experimental studies carried out by Frederic Bartlett in the 1930s. Bartlett showed in these, very clearly (and sometimes very entertainingly), how with retelling a story—either to others or to oneself—the memory of it is continually changed. There was never, Bartlett felt, a simple mechanical reproduction in memory; it was always an individual and imaginative reconstruction. He wrote,

> Remembering is not the re-excitation of innumerable fixed, lifeless and fragmentary traces. It is an imaginative reconstruction, or construction, built out of the relation of our attitude towards a whole active mass of organized past reactions or experience, and to a little outstanding detail which com-

> monly appears in image or in language form. It is thus hardly ever really exact, even in the most rudimentary cases of rote recapitulation, and it is not at all important that it should be so.

Since the last third of the twentieth century, the whole tenor of neurology and neuroscience has been moving towards such a dynamic and constructional view of the brain, a sense that even at the most elementary levels—as, for example, in the "filling in" of a blind spot or scotoma or the seeing of a visual illusion, as both Richard Gregory and V. S. Ramachandran have demonstrated—the brain constructs a plausible hypothesis or pattern or scene. In his theory of neuronal group selection, Gerald Edelman—drawing on the data of neuroanatomy and neurophysiology, of embryology and evolutionary biology, of clinical and experimental work, and of synthetic neural modeling—proposes a detailed neurobiological model of the mind in which the brain's central role is precisely that of constructing categories—first perceptual, then conceptual—and of an ascending process, a "bootstrapping," where through repeating recategorization at higher and higher levels, consciousness is finally achieved. Thus, for Edelman, every perception is a creation and every memory a re-creation or recategorization.

Such categories, he feels, depend on the "values" of the organism, those biases or dispositions (partly innate, partly learned) which, for Freud, were characterized as "drives," "instincts," and "affects." The attunement here between Freud's views and Edelman's is striking; here, at least, one has the sense that psychoanalysis and neurobiology can be fully at home with one another, congruent and mutually supportive. And it may be that in this equation of *Nachträglichkeit* with "recategorization" we see a hint of how the two seemingly disparate universes—the universes of human meaning and of natural science—may come together.

The Fallibility of Memory

In 1993, approaching my sixtieth birthday, I started to experience a curious phenomenon—the spontaneous, unsolicited rising of early memories into my mind, memories that had lain dormant for upwards of fifty years. Not merely memories, but frames of mind, thoughts, atmospheres, and passions associated with them—memories, especially, of my boyhood in London before the Second World War. Moved by these, I wrote two short memoirs: one about the grand science museums in South Kensington, which were so much more important than school to me when I was growing up; the other about Humphry Davy, an early-nineteenth-century chemist who had been a hero of mine in those far-off days and whose vividly described experiments excited me and inspired me to emulation. I think a more general autobiographical impulse was stimulated, rather than sated, by these brief writings, and late in 1997 I launched on a three-year project of dredging, reclaiming memories, reconstructing, refining, seeking for unity and meaning, which finally became my book *Uncle Tungsten.*

I expected some deficiencies of memory, partly because the events I was writing of had occurred fifty or more years earlier and most of those who might have shared their memories, or checked my facts, were now dead. And partly because, in writing about the earliest years of my life, I could not call on the letters and journals I later started to keep from the age of eighteen or so.

I accepted that I must have forgotten or lost a great deal but assumed that the memories I did have—especially those which were very vivid, concrete, and circumstantial—were essentially valid and reliable, and it was a shock to me when I found that some of them were not.

A striking example of this, the first that came to my notice, arose in relation to the two bomb incidents that I described in *Uncle Tungsten,* both of which occurred in the winter of 1940–41, when London was bombarded in the Blitz:

> One night, a thousand-pound bomb fell into the garden next to ours, but fortunately it failed to explode. All of us, the entire street, it seemed, crept away that night (my family to a cousin's flat)—many of us in our pajamas—walking as softly as we could (might vibration set the thing off?). The streets were pitch dark, for the blackout was in force, and we all carried electric torches dimmed with red crêpe paper. We had no idea if our houses would still be standing in the morning.

> On another occasion, an incendiary bomb, a thermite bomb, fell behind our house and burned with a terrible, white-hot heat. My father had a stirrup pump, and my brothers carried pails of water to him, but water seemed useless against this infernal fire—indeed, made it burn even more furiously. There was a vicious hissing and sputtering when the water hit the white-hot metal, and meanwhile the bomb was melting its own casing and throwing blobs and jets of molten metal in all directions.

A few months after the book was published, I spoke of these bombing incidents to my brother Michael. Michael was five years my senior and had been with me at Braefield, the boarding school to which we had been evacuated at the beginning of the war (and in which I was to spend four miserable years, beset by bullying schoolmates and a sadistic headmaster). My brother immediately confirmed the first bombing incident, saying, "I remember it exactly as you described it." But regarding the second bombing, he said, "You never saw it. You weren't there."

I was staggered at Michael's words. How could he dispute a memory I would not hesitate to swear on in a court of law and had never doubted as real?

"What do you mean?" I objected. "I can see it all in my mind's eye now, Pa with his pump, and Marcus and David with their buckets of water. How could I see it so clearly if I wasn't there?"

"You never saw it," Michael repeated. "We were

both away at Braefield at the time. But David [our older brother] wrote us a letter about it. A very vivid, dramatic letter. You were enthralled by it." Clearly, I had not only been enthralled but must have constructed the scene in my mind, from David's words, and then appropriated it and taken it for a memory of my own.

After Michael said this, I tried to compare the two memories—the primary one, on which the direct stamp of experience was not in doubt, with the constructed, or secondary, one. With the first incident, I could feel myself into the body of the little boy, shivering in his thin pajamas—it was December, and I was terrified—and because of my shortness compared with the big adults all around me, I had to crane my head upwards to see their faces.

The second image, of the thermite bomb, was equally clear, it seemed to me—very vivid, detailed, and concrete. I tried to persuade myself that it had a different quality from the first, that it bore evidences of its appropriation from someone else's experience and its translation from verbal description into image. But although I knew, intellectually, that this memory was false, it still seemed to me as real, as intensely my own, as before.[1] Had it, I wondered, become as real, as personal, as strongly embedded

1. On further reflection, I am struck by the way in which I could visualize the garden scene from different angles, whereas the street scene is always "seen" through the eyes of the frightened seven-year-old I was in 1940.

in my psyche (and, presumably, my nervous system) as if it had been a genuine primary memory? Would psychoanalysis, or, for that matter, brain imaging, be able to tell the difference?

~~~~

My false bomb experience was closely akin to the true one, and it could easily have been my own experience, had I been home from school at the time. I could imagine every detail of the garden I knew so well. Had this not been the case, perhaps the description of it in my brother's letter would not have affected me so. But since I could easily imagine being there, and the feelings that would go with this, I took it as my own.

All of us transfer experiences to some extent, and at times we are not sure whether an experience was something we were told or read about, even dreamed about, or something that actually happened to us. This is especially apt to happen with one's so-called earliest memories.

I have a vivid memory from about the age of two of pulling the tail of our chow, Peter, while he was gnawing a bone under the hall table, of Peter leaping up and biting me in the cheek, and of my being carried, howling, into my father's surgery in the house, where a couple of stitches were put in my cheek. There is at least
~~~~

an objective reality here: I was bitten on the cheek by Peter when I was two and still bear the scar of this. But do I actually remember it, or was I told about it, subsequently constructing a "memory" which became more and more firmly fixed in my mind by repetition? The memory seems intensely real to me, and the fear associated with it is certainly real, for I developed a fear of large animals after this incident—Peter was almost as large as I was at two—a fear that they would suddenly attack or bite me.

Daniel Schacter has written extensively on distortions of memory and the source confusions that go with them, and in his book *Searching for Memory* he recounts a well-known story about Ronald Reagan:

> In the 1980 presidential campaign, Ronald Reagan repeatedly told a heartbreaking story of a World War II bomber pilot who ordered his crew to bail out after his plane had been seriously damaged by an enemy hit. His young belly gunner was wounded so seriously that he was unable to evacuate the bomber. Reagan could barely hold back his tears as he uttered the pilot's heroic response: "Never mind. We'll ride it down together." The press soon realized that this story was an almost exact duplicate of a scene in the 1944 film *A Wing and a Prayer.* Reagan had apparently retained the facts but forgotten their source.

Reagan was a vigorous sixty-nine-year-old at the time, would go on to be president for eight years, and only developed unmistakable dementia in his eighties. But he had been given to acting and make-believe throughout his life and had long displayed a vein of romantic fantasy and histrionism. Reagan was not simulating emotion when he recounted this story—his story, his reality, as he felt it to be—and had he taken a lie detector test (functional brain imaging had not yet been invented at the time), there would have been none of the telltale reactions that go with conscious falsehood, for he *believed* what he was saying.

It is startling to realize, though, that some of our most cherished memories may never have happened—or may have happened to someone else.

I suspect that many of my own enthusiasms and impulses, which seem entirely my own, may have arisen from others' suggestions that have powerfully influenced me, consciously or unconsciously, and then been forgotten.

Similarly, while I often give lectures on certain topics, I can never remember, for better or worse, exactly what I said on previous occasions; nor can I bear to look through my earlier notes (or often, even the notes I have made for the talk an hour earlier). Losing conscious memory of what I have said before, I discover my themes afresh each time.

These forgettings may sometimes extend to auto-plagiarism, where I find myself reproducing entire phrases or sentences as if new, and this may be compounded, occasionally, by a genuine forgetfulness.

Looking back through my old notebooks, I find that many of the thoughts sketched in them are forgotten for years, and then revived and reworked as new. I suspect that such forgettings occur for everyone, and they may be especially common in those who write or paint or compose, for creativity may require such forgettings, in order that one's memories and ideas can be born again and seen in new contexts and perspectives.

~~~~

W*ebster's* defines "plagiarize" as "to steal and pass off as one's own the ideas or words of another; use . . . without crediting the source . . . to commit literary theft; present as new and original an idea or product derived from an existing source." There is a considerable overlap between this definition and that of cryptomnesia, and the essential difference is this: plagiarism, as
~~~~

commonly understood and reprobated, is conscious and intentional, whereas cryptomnesia is neither. Perhaps the term "cryptomnesia" needs to be better known, for though one may speak of "unconscious plagiarism," the very word "plagiarism" is so morally charged, so suggestive of crime and deceit, that it retains a sting even if it is unconscious.

In 1970, George Harrison released an enormously successful song, "My Sweet Lord," which turned out to have strong similarities to a song by Ronald Mack ("He's So Fine"), recorded eight years earlier. When the matter went to trial, the court found Harrison guilty of plagiarism, but showed a great deal of psychological insight and sympathy in its judgment. The judge concluded,

> Did Harrison deliberately use the music of "He's So Fine"? I do not believe he did so deliberately. Nevertheless . . . this is, under the law, infringement of copyright, and is no less so though subconsciously accomplished.

Helen Keller was also accused of plagiarism, when she was only twelve.[2] Though deaf and blind from an early age and indeed languageless before she met Annie Sullivan at the age of six, Helen became a prolific writer once

2. This episode is related in great and sympathetic detail by Dorothy Herrmann in her biography of Keller.

she learned finger spelling and Braille. She wrote, among other things, a story called "The Frost King," which she gave to a friend as a birthday gift. When the story found its way into print in a magazine, readers soon realized that it bore great similarities to "The Frost Fairies," a children's short story by Margaret Canby. Admiration for Keller turned into condemnation, and she was accused of plagiarism and deliberate falsehood, even though she had no recollection of reading Mrs. Canby's story. (She later realized that the story had been "read" to her, using finger spelling onto her hand.) The young Keller was subjected to a ruthless and outrageous inquisition, which left its mark on her for the rest of her life.

But she had defenders, too, including the plagiarized Margaret Canby, who was amazed that a story spelled into Keller's hand three years before could be remembered or reconstructed by her in such detail. "What a wonderfully active and retentive mind that gifted child must have!" Canby wrote. Alexander Graham Bell, too, came to her defense, saying, "Our most original compositions are composed exclusively of expressions derived from others."

Keller herself later said of such appropriations that they were most apt to occur when books were spelled into her hands, their words passively received. Sometimes when this was done, she said, she could not identify or remember their source, nor even, sometimes, whether they came from outside her or not. Such confusion rarely occurred if she read actively, using Braille, moving her finger across the pages.

Mark Twain wrote, in a letter to Keller,

> Oh, dear me, how unspeakably funny and owlishly idiotic and grotesque was that "plagiarism" farce! As if there was much of anything in any human utterance, oral or written, *except* plagiarism! . . . For substantially all ideas are second-hand, consciously and unconsciously drawn from a million outside sources.

Indeed, Twain had committed such unconscious theft himself, as he described in a speech at Oliver Wendell Holmes's seventieth birthday:

> Oliver Wendell Holmes [was] the first great literary man I ever stole any thing from—and that is how I came to write to him and he to me. When my first book was new, a friend of mine said to me, "The dedication is very neat." Yes, I said, I thought it was. My friend said, "I always admired it, even before I saw it in *The Innocents Abroad*."
>
> I naturally said, "What do you mean? Where did you ever see it before?"
>
> "Well, I saw it first some years ago as Doctor Holmes's dedication to his *Songs in Many Keys*."
>
> Of course, my first impulse was to prepare this

man's remains for burial, but upon reflection I said I would reprieve him for a moment or two and give him a chance to prove his assertion if he could: We stepped into a book-store, and he did prove it. I had really stolen that dedication, almost word for word. . . .

Well, of course, I wrote to Doctor Holmes and told him I hadn't meant to steal, and he wrote back and said in the kindest way that it was all right and no harm done; and added that he believed we all unconsciously worked over ideas gathered in reading and hearing, imagining that they were original with ourselves.

He stated a truth, and did it in such a pleasant way . . . that I was rather glad I had committed the crime, for the sake of the letter. I afterwards called on him and told him to make perfectly free with any ideas of mine that struck him as being good protoplasm for poetry. He could see by that that there wasn't anything mean about me; so we got along right from the start.

~~~~

The question of Coleridge's plagiarisms, paraphrases, cryptomnesias, or borrowings has intrigued scholars and biographers for nearly two centuries and is of special
~~~~

interest in view of his prodigious powers of memory, his imaginative genius, and his complex, multiform, sometimes tormented sense of identity. No one has described this more beautifully than Richard Holmes in his two-volume biography.

Coleridge was a voracious, omnivorous reader, who seemed to retain all that he read. There are descriptions of him as a student reading *The Times* in a casual fashion, then being able to reproduce the entire paper, including its advertisements, verbatim. "In the youthful Coleridge," writes Holmes, "this is really part of his gift: an enormous reading capacity, a retentive memory, a talker's talent for conjuring and orchestrating other people's ideas, and the natural instincts of a lecturer and preacher to harvest materials wherever he found them."

Literary borrowing was commonplace in the seventeenth century; Shakespeare borrowed freely from many of his contemporaries, as did Milton. Friendly borrowing remained common in the eighteenth century, and Coleridge, Wordsworth, and Southey all borrowed from each other, sometimes even, according to Holmes, publishing work under each other's names.

But what was common, natural, and playful in Coleridge's youth gradually took on a more disquieting form, especially in relation to the German philosophers (Friedrich Schelling above all) whom he discovered, venerated, and translated into English. Whole pages of Coleridge's *Biographia Literaria* consist of unacknowledged, verbatim passages from Schelling. While this unconcealed and damaging behavior has been readily (and reductively) categorized as "literary kleptomania," what actually went on is complex and mysterious, as Holmes explores in the second volume of his biography, where he sees the most flagrant of Coleridge's plagiarisms as occurring at a devastatingly difficult period of his life, when he had been abandoned by Wordsworth, was disabled by profound anxiety and intellectual self-doubt, and was more deeply addicted to opium than ever. At this time, Holmes writes, "his German authors gave him support and comfort: in a metaphor he often used himself, he twined round them like ivy round an oak."

Earlier, as Holmes describes, Coleridge had found another extraordinary affinity, with the German writer Jean Paul Richter—an affinity which led him to translate Richter's writings and then to take off from them, elaborating them in his own way, conversing and communing in his notebooks with Richter. At times, the voices of the two men became so intermingled as to be hardly distinguishable from each other.

~~~~

In 1996, I read a review of a new play, *Molly Sweeney,* by the eminent playwright Brian Friel. His lead character, Molly, I read, had been born blind but has her sight restored in middle age. She can see clearly after her opera-
~~~~

tion, yet she can recognize nothing: she has visual agnosia because her brain has never learned to see. She finds this frightening and bizarre and is relieved when she returns to her original state of blindness. I was startled by this, because I had published an exceedingly similar story in *The New Yorker* only three years earlier.[3] Indeed, when I read Friel's play, I was surprised to find, over and above the thematic similarities, a great many phrases and sentences from my own case history. When I contacted Friel to ask him about this, he denied even knowing about my essay—but then, after I sent him a detailed comparison of the two, he realized that he must have read my piece but forgotten doing so. He was confounded: he had read many of the same original sources I mentioned in my article, and believed that the themes and language of *Molly Sweeney* were entirely original. Somehow, he concluded, he had unconsciously absorbed much of my own language, thinking it was his own. (He agreed to add an acknowledgment of this to the play.)

Freud was fascinated by the slippages and errors of memory that occur in the course of daily life and

3. This essay, "To See and Not See," was subsequently published in my book *An Anthropologist on Mars*.

their relation to emotion, especially unconscious emotion. But he was also forced to consider the much grosser distortions of memory that some of his patients showed, especially when they gave him accounts of having been sexually seduced or abused in childhood. At first he took all these accounts literally, but eventually, when there seemed little evidence or plausibility in several cases, he started to wonder whether such recollections had been distorted by fantasy and whether some, indeed, might be total fabulations, constructed unconsciously, but so convincingly that the patients themselves believed in them absolutely. The stories that patients told, and had told to themselves, even if they were false, could have a very powerful effect on their lives, and it seemed to Freud that their psychological reality might be the same whether they came from actual experience or from fantasy.

In a 1995 memoir, *Fragments,* Binjamin Wilkomirski described how, as a Polish Jew, he had spent several years of his childhood surviving the horrors and dangers of a concentration camp. The book was hailed as a masterpiece. A few years later, it was discovered that Wilkomirski had been born not in Poland but in Switzerland, was not Jewish, and had never been in a concentration camp. The entire book was an extended fabulation. (Elena Lappin described this in a 1999 essay in *Granta.*)

While there were outraged accusations of fraud, it seemed, on further exploration, that Wilkomirski had not intended to deceive his readers (nor, indeed, had he originally wanted the book to be published). He had, for many years, been engaged in an enterprise of his own—basically the romantic reinvention of his own childhood, apparently in reaction to his abandonment by his mother at the age of seven.

Apparently, Wilkomirski's primary intention was to deceive himself. When he was confronted with the actual historical reality, his reaction was one of bewilderment and confusion. He was totally lost, by this point, in his own fiction.

Much is made of so-called recovered memories—memories of experiences so traumatic as to be defensively repressed and then, with therapy, released from repression. Particularly dark and fantastic forms of this include descriptions of satanic rituals of one sort or another, accompanied often by coercive sexual practices. Lives, and families, have been ruined by such accusations. But it has been shown that such descriptions, in at least some cases, can be insinuated or planted by others. The frequent combination of a suggestible witness (often a child) with an authority figure (perhaps a therapist, a teacher, a social worker, or an investigator) can be particularly powerful.

From the Inquisition and the Salem witch trials to the Soviet trials of the 1930s and Abu Ghraib, varieties of "extreme interrogation," or outright physical and mental torture, have been used to extract religious or political "confessions." While such interrogation may be designed to extract information in the first place, its deeper intentions may be to brainwash, to effect a genuine change of mind, to fill it with implanted, self-inculpatory memories—and in this it may be frighteningly successful. (There is no parable more relevant here than Orwell's *1984,* where Winston, at the end, under unbearable pressure, is finally broken, betrays Julia, betrays himself and all his ideals, betrays his memory and judgment, too, and ends up loving Big Brother.)

But it may not take massive or coercive suggestion to affect a person's memories. The testimony of eyewitnesses is notoriously subject to suggestion and to error, frequently with dire results for the wrongfully accused. With DNA testing, it is now possible to find, in many cases, an objective corroboration or refutation of such testimony, and Schacter has noted that "a recent analysis of forty cases in which DNA evidence established the innocence of wrongly imprisoned individuals revealed that thirty-six of them (90 percent) involved mistaken eyewitness identification."[4]

4. Hitchcock's film *The Wrong Man* (the only nonfiction film he ever made) documents the terrifying consequences of a mistaken identification based on eyewitness testimony ("leading" the witnesses, as well as accidental resemblances, plays a major part here).

If the last few decades have seen a surge or resurgence of ambiguous memory and identity syndromes, they have also led to important research—forensic, theoretical, and experimental—on the malleability of memory. Elizabeth Loftus, the psychologist and memory researcher, has documented a disquieting success in implanting false memories by simply suggesting to a subject that he has experienced a fictitious event. Such pseudo-events, invented by psychologists, may vary from comic incidents to mildly upsetting ones (for example, that one was lost in a shopping mall as a child) to more serious incidents (that one was the victim of an animal attack or an assault by another child). After initial skepticism ("I was never lost in a shopping mall") and then uncertainty, the subject may move to a conviction so profound that he will continue to insist on the truth of the implanted memory even after the experimenter confesses that it never happened in the first place.

What is clear in all these cases—whether of imagined or real abuse in childhood, of genuine or experimentally implanted memories, of misled witnesses and brainwashed prisoners, of unconscious plagiarism, and of the false memories we all have based on misattribution or source confusion—is that in the absence of outside confirmation there is no easy way of distinguishing a genuine memory or inspiration, felt as such, from those that have been borrowed or suggested, between what Donald Spence calls "historical truth" and "narrative truth."

Even if the underlying mechanism of a false memory

is exposed, as I was able to do, with my brother's help, in the incendiary bomb incident (or as Loftus would do when she confessed to her subjects that their memories were implanted), this may not alter the sense of actual lived experience or "reality" which such memories have. Nor, for that matter, may the obvious contradictions or absurdity of certain memories alter the sense of conviction or belief. For the most part, people who claim to have been abducted by aliens are not lying when they speak of their experiences, any more than they are conscious of having invented a story—they truly believe that this is what happened. (In *Hallucinations,* I describe how hallucinations, whether caused by sensory deprivation, exhaustion, or various medical conditions, can be taken as reality partly because they involve the same sensory pathways in the brain that "real" perceptions do.)

Once such a story or memory is constructed, accompanied by vivid sensory imagery and strong emotion, there may be no inner, psychological way of distinguishing true from false, nor any outer, neurological way. The physiological correlates of such memories can be examined using functional brain imaging, and these images show that vivid memories produce widespread activation in the brain involving sensory areas, emotional (limbic) areas, and executive (frontal lobe) areas—a pattern that is virtually identical whether the "memory" is based on experience or not.

There is, it seems, no mechanism in the mind or

the brain for ensuring the truth, or at least the veridical character, of our recollections. We have no direct access to historical truth, and what we feel or assert to be true (as Helen Keller was in a very good position to note) depends as much on our imagination as our senses. There is no way by which the events of the world can be directly transmitted or recorded in our brains; they are experienced and constructed in a highly subjective way, which is different in every individual to begin with, and differently reinterpreted or reexperienced whenever they are recollected. Our only truth is narrative truth, the stories we tell each other and ourselves—the stories we continually recategorize and refine. Such subjectivity is built into the very nature of memory and follows from its basis and mechanisms in the brains we have. The wonder is that aberrations of a gross sort are relatively rare and that for the most part our memories are so solid and reliable.

We, as human beings, are landed with memories which have fallibilities, frailties, and imperfections—but also great flexibility and creativity. Confusion over sources or indifference to them can be a paradoxical strength: if we could tag the sources of all our knowledge, we would be overwhelmed with often irrelevant information. Indifference to source allows us to assimilate what we read, what we are told, what others say and think and write and paint, as intensely and richly as if they were primary experiences. It allows us to see and hear with other eyes and ears, to enter into other minds, to assimilate the art and science and religion of the whole culture, to enter into and contribute to the common mind, the general commonwealth of knowledge. Memory arises not only from experience but from the intercourse of many minds.

Mishearings

A few weeks ago, when I heard my friend Kate say to me, "I am going to choir practice," I was surprised. I have never, in the thirty years we have known each other, heard her express the slightest interest in singing. But I thought, who knows? Perhaps this is a part of herself she has kept quiet about; perhaps it is a new interest; perhaps her son is in a choir; perhaps . . .

I was fertile with hypotheses, but I did not consider for a moment that I had misheard her. It was only on her return that I found she had been to the *chiropractor.*

A few days later, Kate jokingly said, "I'm off to choir practice." Again I was baffled: Firecrackers? Why was she talking about firecrackers?

As my deafness increases, I am more and more prone to mishearing what people say, though this is quite unpredictable; it may happen twenty times, or not at all, in the course of a day. I carefully record these in a little red notebook labeled PARACUSES—alterations in hearing, especially mishearings. I enter what I hear (in red) on one page, what was actually said (in green) on the opposite page, and (in purple) people's reactions to my mishearings and the often far-fetched hypotheses I may consider in an attempt to make sense of what is often essentially nonsensical.

After the publication of Freud's *Psychopathology of Everyday Life* in 1901, such mishearings, along with a range of misreadings, misspeakings, misdoings, and slips of the tongue, were seen as "Freudian," an expression of deeply repressed feelings and conflicts. But although there are occasional, unprintable mishearings that make me blush, a vast majority do not admit any simple Freudian interpretation. In almost all of my mishearings, however, there is a similar overall sound, a similar acoustic gestalt, linking what is said and what is heard. Syntax is always preserved, but this does not help; mishearings are likely to capsize meaning, to overwhelm it with phonologically similar but meaningless or absurd sound forms, even though the general form of a sentence is preserved.

Lack of clear enunciation, unusual accents, or poor transmission can all serve to mislead one's own perceptions. Most mishearings substitute one real word for another, however absurd or out of context, but sometimes the brain comes up with a neologism. When a friend told me on the phone that her child was sick, I misheard "tonsillitis" as "pontillitis," and I was puzzled. Was this some unusual clinical syndrome, an inflammation I had never heard of? It did not occur to me that I had invented a nonexistent word—indeed, a nonexistent condition.

Every mishearing is a novel concoction. The hundredth mishearing is as fresh and as surprising as the first. I am often strangely slow to realize that I have misheard, and I may entertain the most elaborate ideas to explain my mishearings, when it would seem that I should spot them straightaway. If a mishearing seems plausible, one may not think one *has* misheard; it is only if the mishearing is sufficiently implausible, or entirely out of context, that one thinks, "This can't be right," and (perhaps with some embarrassment) asks the speaker to repeat himself, as I often do, or even to spell out misheard words or phrases.

When Kate spoke of going to choir practice, I accepted this: she *could* have been going to choir practice. But when a friend spoke one day about "a big-time cuttlefish diagnosed with ALS," I felt I *must* be mishearing. Cephalopods have elaborate nervous systems, it is true, and perhaps, I thought for a split second, a cuttlefish *could* have ALS. But the idea of a "big-time" cuttlefish was ridiculous. (It turned out to be "a big-time publicist.")

While mishearings may seem to be of little special interest, they can cast an unexpected light on the nature of perception—the perception of speech, in particular. What is extraordinary, first, is that they present themselves as clearly articulated words or phrases, not as jumbles of sound. One *mishears* rather than just fails to hear.

Mishearings are not hallucinations, but like hallucinations they utilize the usual pathways of perception and pose as reality—it does not occur to one to question them. But since all of our perceptions must be constructed by the brain from often meager and ambiguous sensory data, the possibility of an error or deception is always present. Indeed, it is a marvel that our perceptions are so often correct, given the rapidity, the near instantaneity, with which they are constructed.

One's surroundings, one's wishes and expectations, conscious and unconscious, can certainly be codeterminants in mishearing, but the real mischief lies at lower levels, in those parts of the brain involved in phonological analysis and decoding. Doing what they can with distorted or deficient signals from our ears, these parts of the brain manage to construct real words or phrases, even if they are absurd.

While I often mishear words, I seldom mishear music: notes, melodies, harmonies, phrasings, remain as clear and rich as they have been all my life (though I often mishear *lyrics*). There is clearly something about the way the brain processes music that makes it robust, even in the face of imperfect hearing, and, conversely, something about the nature of spoken language that makes it much more vulnerable to deficiencies or distortions.

Playing or even hearing music (at least traditional scored music) involves not just the analysis of tone and rhythm; it also engages one's procedural memory and emotional centers in the brain; musical pieces are held in memory and allow anticipation.

But speech must be decoded by other systems in the brain as well, including systems for semantic memory and syntax. Speech is open, inventive, improvised; it is rich in ambiguity and meaning. There is a huge freedom in this, making spoken language almost infinitely flexible and adaptable but also vulnerable to mishearing.

Was Freud entirely wrong, then, about slips and mishearings? Of course not. He advanced fundamental considerations about wishes, fears, motives, and conflicts not present in consciousness, or thrust out of consciousness, which could color slips of the tongue, mishearings, or misreadings. But he was perhaps too insistent that misperceptions are wholly a result of unconscious motivation.

Collecting mishearings over the past few years without any explicit selection or bias, I am forced to think that Freud underestimated the power of neural mechanisms, combined with the open and unpredictable nature of language, to sabotage meaning, to generate mishearings that are irrelevant both in terms of context *and* of subconscious motivation.

And yet there is often a sort of style or wit—a "dash"—in these instantaneous inventions; they reflect, to some extent, one's own interests and experiences, and I rather enjoy them. Only in the realm of mishearing—at least my mishearings—can a biography of cancer become a biography of Cantor (one of my favorite mathematicians), tarot cards turn into pteropods, a grocery bag into a poetry bag, all-or-noneness into oral numbness, a porch into a Porsche, and a mere mention of Christmas Eve a command to "Kiss my feet!"

The Creative Self

All children indulge in play, at once repetitive and imitative and, equally, exploratory and innovative. They are drawn both to the familiar and the unusual—grounding and anchoring themselves in what is known and secure, and exploring what is new and has never been experienced. Children have an elemental hunger for knowledge and understanding, for mental food and stimulation. They do not need to be told or "motivated" to explore or play, for play, like all creative or proto-creative activities, is deeply pleasurable in itself.

Both the innovative and the imitative impulses come together in pretend play, often using toys or dolls or miniature replicas of real-world objects to act out new scenarios or rehearse and replay old ones. Children are drawn to narrative, not only soliciting and enjoying stories from others, but creating them themselves. Storytelling and mythmaking are primary human activities, a fundamental way of making sense of our world.

Intelligence, imagination, talent, and creativity will get nowhere without a basis of knowledge and skills, and for this education must be sufficiently structured and focused. But an education too rigid, too formulaic, too lacking in narrative, may kill the once-active, inquisitive mind of a child. Education has to achieve a balance between structure and freedom, and each child's needs may be extremely variable. Some young minds expand and blossom with good teaching. Other children (including some of the most creative) may be resistant to formal teaching; they are essentially autodidacts, voracious to learn and explore on their own. Most children will go through many stages in this process, needing more or less structure, more or less freedom at different periods.

Voracious assimilation, imitating various models, while not creative in itself, is often the harbinger of future creativity. Art, music, film, and literature, no less than facts and information, can provide a special sort of education, what Arnold Weinstein calls a "vicarious immersion in others' lives, endowing us with new eyes and ears."

For my generation, this immersion came mostly from reading. Susan Sontag, at a conference in 2002, spoke about how reading opened up the entire world to her when she was quite young, enlarging her imagination and memory far beyond the bounds of her actual, immediate personal experience. She recalled,

> When I was five or six, I read Eve Curie's biography of her mother. I read comic books, dictionar-

> ies, and encyclopedias indiscriminately, and with great pleasure. . . . It felt like the more I took in, the stronger I was, the bigger the world got. . . . I think I was, from the very beginning, an incredibly gifted student, an incredibly gifted learner, a champion child autodidact. . . . Is that creative? No, it wasn't creative . . . [but] it didn't preclude becoming creative later on. . . . I was engorging rather than making. I was a mental traveler, a mental glutton. . . . My childhood, apart from my wretched actual life, was just a career in ecstasy.

What is especially striking about Sontag's account (and similar accounts of proto-creativity) is the energy, the ravenous passion, the enthusiasm, the love with which the young mind turns to whatever will nourish it, seeks intellectual or other models, and hones its skills by imitation.

She assimilated a vast knowledge of other times and places, of the varieties of human nature and experience, and these perspectives played a huge part in inciting her to write herself:

> I started writing when I was about seven. I started a newspaper when I was eight, which I filled with stories and poems and plays and articles, and which I used to sell to the neighbors for five cents. I'm sure it was quite banal and conventional, and simply made

> up of things, influenced by things, I was reading. . . . Of course there were models, there was a pantheon of these people. . . . If I was reading the stories of Poe, then I would write a Poe-like story. . . . When I was ten, a long-forgotten play by Karel Čapek, *R.U.R.*, about robots, fell into my hands, so I wrote a play about robots. But it was absolutely derivative. Whatever I saw I loved, and whatever I loved I wanted to imitate—that's not necessarily the royal road to real innovation or creativity; neither, as I saw it, does it preclude it. . . . I started to be a real writer at thirteen.

Sontag's prodigious and precocious intelligence and creativity allowed her to leapfrog into "real" writing as a teenager, but for most people the period of imitation and apprenticeship, studenthood, lasts much longer. It is a time when one struggles to find one's own powers, one's own voice. It is a time of practice, repetition, of mastering and perfecting skills and techniques.

Some people, having undergone such an apprenticeship, may remain at the level of technical mastery without ever ascending to major creativity. And it may be difficult to judge, even at a distance, when the leap from talented but derivative work to major innovation has occurred. Where does one draw the line between influence and imitation? What distinguishes a creative assimilation, a deep intertwining of appropriation and experience, from mere mimicry?

~~~~

The term "mimicry" may imply a certain consciousness or intention, but imitating, echoing, or mirroring are universal psychological (and indeed physiological) propensities that can be seen in every human being, and many animals (hence terms like "parroting" or "aping"). If one sticks one's tongue out at a young infant, it will mirror this behavior, even before it has gained adequate control of its limbs, or has much of a body image—and such mirroring remains an important mode of learning throughout life.

Merlin Donald, in his book *Origins of the Modern Mind,* sees "mimetic culture" as a crucial stage in the evolution of culture and cognition. He draws a clear distinction between mimicry, imitation, and mimesis:

> Mimicry is literal, an attempt to render as exact a duplicate as possible. Thus, exact reproduction of a facial expression, or exact duplication of the sound of another bird by a parrot, would constitute mimicry. . . . Imitation is not so literal as mimicry; the offspring copying its parent's behavior imitates, but does not mimic, the parent's way of doing things. . . . Mimesis adds a representational dimension to imitation. It usually incorporates both mimicry and imitation to a higher end, that of re-enacting and re-presenting an event or relationship.
~~~~

Mimicry, Donald suggests, occurs in many animals; imitation in monkeys and apes; mimesis solely in humans. But all can coexist and overlap in us—a performance, a production, may have elements of all three.

In certain neurological conditions the powers of mimicry and reproduction may be exaggerated, or perhaps less inhibited. People with Tourette's syndrome or autism or certain types of frontal lobe damage, for example, may be unable to inhibit an involuntary echoing or mirroring of other people's speech or actions; they may also echo sounds, even meaningless sounds in the environment. In *The Man Who Mistook His Wife for a Hat,* I describe one woman with Tourette's who, while walking down the street, would echo or imitate the "dental" grilles of cars, the gibbet-like forms of lampposts, and the gestures and walks of everyone she passed—often exaggerated in a caricature-like way.

Some autistic savants have very exceptional powers of visual imagery and reproduction. This is evident with Stephen Wiltshire, whom I describe in *An Anthropologist on Mars.* Stephen is a visual savant with a great gift for catching visual likenesses. It makes little difference whether he makes these from life, on the spot, or long afterwards—perception and memory, here, seem almost indistinguishable. He also has an amazing ear; when he was a child, he would echo noises and words, seemingly without any intention or any consciousness of it. When, as an adolescent, he returned from a visit to Japan, he kept emitting "Japanese" noises, babbling pseudo-Japanese, and showing "Japanese" gestures, too. He can imitate the sound of any musical instrument once he has heard it and has a very accurate musical memory. I was very struck once when, at sixteen, he sang and mimed Tom Jones's song "It's Not Unusual," swinging his hips, dancing, gesticulating, and clutching an imaginary microphone to his mouth. At this age, Stephen usually displayed very little emotion and showed many of the outward manifestations of classical autism, with a skewed neck posture, tics, and indirection of gaze, but all this disappeared when he sang the Tom Jones song—so much so that I wondered whether, in some uncanny way, he had gone beyond mimicry and actually shared the emotion and sensibility of the song. I was reminded of an autistic boy I had met in Canada who knew an entire television show by heart and would "replay" this dozens of times a day, complete with all the voices and gestures, and even the sounds of applause. I felt this as a sort of automatism or superficial reproduction, but Stephen's performance left me puzzled and pensive. Had he, unlike the Canadian boy, moved from mimicry into creativity or art? Was he consciously and intentionally sharing the emotions and sensibility of the song or merely reproducing it—or something in between?[1]

1. In autistic or retarded people with savant syndrome, the powers of retention and reproduction may be prodigious, but what is retained

Another autistic savant, José (whom I also describe in *The Man Who Mistook His Wife for a Hat*), was often described by the hospital staff as a sort of copying machine. This was unfair and insulting, as well as incorrect, for the retentiveness of a savant's memory is not at all comparable to a mechanical process; there is discrimination and recognition of visual features, speech features, particularities of gesture, etc. But to some extent, the "meaning" of these is not fully incorporated, and this makes the savant memory, to the rest of us, seem comparatively mechanical.

~~~~

If imitation plays a central role in the performing arts, where incessant practice, repetition, and rehearsal are essential, it is equally important in painting or compos-

---

is apt to be retained as something external, indifferently. Langdon Down, who identified Down's syndrome in 1862, wrote about one savant boy who "having once read a book, could ever more remember it." Once Down gave the boy a copy of Gibbon's *Decline and Fall* to read. The boy read and recited this fluently but without any comprehension, and on page 3 he skipped a line but then went back and corrected himself. "Ever after," Down wrote, "when reciting from memory the stately periods of Gibbon, he would, on coming to the third page, skip the line and go back and correct the error with as much regularity as if it had been part of the regular text."
~~~~

ing or writing, for example. All young artists seek models in their apprentice years, models whose style, technical mastery, and innovations can teach them. Young painters may haunt the galleries of the Met or the Louvre; young composers may go to concerts or study scores. All art, in this sense, starts out as "derivative," highly influenced by, if not a direct imitation or paraphrase of, the admired and emulated models.

When Alexander Pope was thirteen years old, he asked William Walsh, an older poet whom he admired, for advice. Walsh's advice was that Pope should be "correct." Pope took this to mean that he should first gain a mastery of poetic forms and techniques. To this end, in his "Imitations of English Poets," Pope began by imitating Walsh, then Cowley, the Earl of Rochester, and more major figures like Chaucer and Spenser, as well as writing "Paraphrases," as he called them, of Latin poets. By seventeen, he had mastered the heroic couplet and began to write his "Pastorals" and other poems, where he developed and honed his own style but contented himself with the most insipid or clichéd themes. It was only once he had established full mastery of his style and form that he started to charge it with the exquisite and sometimes terrifying products of his own imagination. For most artists, perhaps, these stages or processes overlap a good deal, but imitation and mastery of form or skills must come before major creativity.

Yet even with years of preparation and conscious mastery, great talent may or may not fulfill its seeming promise.[2] Many creators—whether they are artists, scientists, cooks, teachers, or engineers—are content, after they have achieved a level of mastery, to stay with a form, or play within its bounds, for the rest of their lives, never breaking into anything radically new. Their work may still show mastery and even virtuosity, giving great delight even if it does not take the further step into "major" creativity.

There are many examples of "minor" creativity, creativity that does not seem to change much in character after its initial expression. Arthur Conan Doyle's 1887 *A Study in Scarlet,* the first of his Sherlock Holmes books, was a remarkable achievement—there had never been a "detective story" like it before.[3] *The Adventures of Sherlock Holmes,* five years later, was an enormous success, and Conan Doyle found himself the acclaimed writer of a potentially never-ending series. He was delighted by this

2. In his autobiography, *Ex-Prodigy,* Norbert Wiener, who entered Harvard at fourteen to get his doctorate and remained a prodigy all his life, describes a contemporary of his, William James Sidis. Sidis (named after his godfather, William James) was a brilliant polyglot mathematician who enrolled at Harvard at the age of eleven but by sixteen, perhaps overwhelmed by the demands of his genius and of society, had given up mathematics and retreated from public and academic life.

3. There had been Poe's Dupin stories ("The Murders in the Rue Morgue," for example), but these had none of the personal quality, the rich characterization of Holmes and Watson.

but also annoyed, because he wanted to write historical novels too, but the public showed little interest in these. They wanted Holmes and more Holmes, and he had to supply it. Even after he killed Holmes off in "The Final Problem," sending him over the Reichenbach Falls in mortal combat with Moriarty, the public insisted that he be resurrected, and so he was, in 1905, in *The Return of Sherlock Holmes.*

There is not much development in Holmes's method or mind or character; he does not seem to age. Between cases, Holmes himself scarcely exists—or, rather, exists in a regressed state: scraping at his violin, shooting up cocaine, doing malodorous chemical experiments—until he is summoned to action by the next case. The stories of the 1920s could have been written in the 1890s, and those written in the 1890s would not have been out of place later. Holmes's London is as unchanging as the man; both are depicted, brilliantly, once and for all, in the 1890s. Doyle himself, in his 1928 preface to *Sherlock Holmes: The Complete Short Stories,* says that the reader may read the stories "in any order."

Why is it that of every hundred gifted young musicians who study at Juilliard or every hundred brilliant young scientists who go to work in major labs under illustrious mentors, only a handful will write memorable musical compositions or make scientific discoveries of major importance? Are the majority, despite their gifts, lacking in some further creative spark? Are they missing characteristics other than creativity that may be essential for creative achievement—such as boldness, confidence, independence of mind?

It takes a special energy, over and above one's creative potential, a special audacity or subversiveness, to strike out in a new direction once one is settled. It is a gamble as all creative projects must be, for the new direction may not turn out to be productive at all.

Creativity involves not only years of conscious preparation and training but unconscious preparation as well. This incubation period is essential to allow the subconscious assimilation and incorporation of one's influences and sources, to reorganize and synthesize them into something of one's own. In Wagner's overture to *Rienzi,* one can almost trace this emergence. There are echoes, imitations, paraphrases, pastiches of Rossini, Meyerbeer, Schumann, and others—all the musical influences of his apprenticeship. And then, suddenly, astoundingly, one hears Wagner's own voice: powerful, extraordinary (though, to my mind, horrible), a voice of genius, without precedent or antecedent. The essential element in these realms of retaining and appropriating versus assimilating and incorporating is one of depth, of meaning, of active and personal involvement.

~~~~

Early in 1982, I received an unexpected packet from London containing a letter from Harold Pinter and the manuscript of a new play, *A Kind of Alaska,* which, he said, had been inspired by a case history of mine in *Awakenings.* In his letter, Pinter said that he had read my book when it originally came out in 1973 and had immediately wondered about the problems presented by a dramatic adaptation of this. But, seeing no ready solution to these problems, he had then forgotten about it. One morning eight years later, Pinter wrote, he had awoken with the first image and first words ("Something is happening") clear and pressing in his mind. The play had then "written itself" in the days and weeks that followed.

I could not help contrasting this with a play (inspired by the same case history) which I had been sent four years earlier, where the author, in an accompanying letter, said that he had read *Awakenings* two months before and been so "influenced," so possessed, by it that he felt impelled to write a play straightaway. Whereas I loved Pinter's play—not least because it effected so profound a transformation, a "Pinterization" of my own
~~~~

themes—I felt the 1978 play to be grossly derivative, for it lifted, sometimes, whole sentences from my own book without transforming them in the least. It seemed to me less an original play than a plagiarism or a parody (yet there was no doubting the author's "obsession" or good faith).

I was not sure what to make of this. Was the author too lazy, or too lacking in talent or originality, to make the needed transformation of my work? Or was the problem essentially one of incubation, that he had not allowed himself enough time for the experience of reading *Awakenings* to sink in? Nor had he allowed himself, as Pinter did, time to forget it, to let it fall into his unconscious, where it might link with other experiences and thoughts.

All of us, to some extent, borrow from others, from the culture around us. Ideas are in the air, and we may appropriate, often without realizing, the phrases and language of the times. We borrow language itself; we did not invent it. We found it, we grew up into it, though we may use it, interpret it, in very individual ways. What is at issue is not the fact of "borrowing" or "imitating," of being "derivative," being "influenced," but what one does with what is borrowed or imitated or derived; how deeply one assimilates it, takes it into oneself, compounds it with one's own experiences and thoughts and feelings, places it in relation to oneself, and expresses it in a new way, one's own.

Time, "forgetting," and incubation are equally necessary before deep scientific or mathematical insights can be achieved. Henri Poincaré, the great mathematician, recounts in his autobiography how he wrestled with a particularly difficult mathematical problem but, getting nowhere, became deeply frustrated.[4] He decided to take a break by going on a geological excursion, and this travel distracted him from his mathematical problem. One day, though, he wrote,

> We entered an omnibus to go some place or other. At the moment when I put my foot on the step, the idea came to me, without anything in my former thoughts seeming to have paved the way for it, that the transformations I had used to define the Fuchsian functions were identical with those of non-Euclidian geometry. I did not verify the idea; I should not have had time, as . . . I went on with a conversation already commenced, but I felt a perfect certainty. On my return to Caen, for conscience's sake, I verified the result at my leisure.

4. Jacques Hadamard relates this in his *Psychology of Invention in the Mathematical Field.*

A while later, "disgusted" with his failure to solve a different problem, he went to the seaside, and here, as he wrote,

> one morning, walking on the bluff, the idea came to me, with just the same characteristics of brevity, suddenness and immediate certainty, that the arithmetic transformations of indefinite ternary quadratic forms were identical with those of non-Euclidian geometry.

It seemed clear, as Poincaré wrote, that there must be an active and intense unconscious (or subconscious, or preconscious) activity even during the period when a problem is lost to conscious thought, and the mind is empty or distracted with other things. This is not the dynamic or "Freudian" unconscious, boiling with repressed fears and desires, nor the "cognitive" unconscious, which enables one to drive a car or to utter a grammatical sentence with no conscious idea of how one does it. It is instead the incubation of hugely complex problems performed by an entire hidden, creative self. Poincaré pays tribute to this unconscious self: it "is not purely automatic; it is capable of discernment; . . . it knows how to choose, to divine. . . . It knows better how to divine than the conscious self, since it succeeds where that has failed."

The sudden uprush of a solution to a long-incubated problem may sometimes occur in dreams, or in states of partial consciousness such as one tends to experience immediately before falling asleep or immediately after waking, with the strange freedom of thought and sometimes almost hallucinatory imagery that can come at such times. Poincaré recorded how one night, while in this sort of twilight state, he seemed to see ideas in motion, like molecules of a gas, occasionally colliding and coupling in pairs, locking together to form more complex ideas—a rare view (though others have described similar ones, especially in drug-induced states) of the usually invisible creative unconscious.

And Wagner gives a vivid description of how the orchestral introduction to *Das Rheingold* came to him, after long waiting, when he too was in a strange, quasi-hallucinatory twilight state:

> After a night spent in fever and sleeplessness, I forced myself to take a long tramp the next day through the hilly country, which was covered with pine-woods. . . . Returning in the afternoon, I stretched myself, dead tired, on a hard couch. . . . I fell into a kind of somnolent state, in which I suddenly felt as though I were sinking in swiftly flowing water. The rushing sound formed itself in my brain into a musical sound, the chord of E flat major, which continued re-echoed in broken forms; these broken forms seemed to be melodic passages of increasing motion, yet the pure triad of E flat major never changed, but

> seemed by its continuance to impart infinite significance to the element in which I was sinking. . . . I at once recognized that the orchestral overture to the Rheingold, which must have long lain latent within me . . . had at last been revealed to me.[5]

Could one distinguish, by some functional brain imaging yet to be invented, the mimicries or imitations of an autistic savant from the deep conscious and the unconscious transformations of a Wagner? Does verbatim memory look different, neurologically, from deep, Proustian memory? Could one demonstrate how some

5. There are many similar stories, some of them iconic, some perhaps mythologized, about sudden scientific discoveries coming in dreams. Mendeleev, the great Russian chemist, discovered the periodic table in a dream, it is said, and, immediately on waking, jotted it down on an envelope. The envelope exists; the account, as far as it goes, may be true. But it gives the impression that this stroke of genius came out of the blue, whereas, in reality, Mendeleev had been pondering the subject, consciously and unconsciously, for at least nine years, ever since the 1860 conference at Karlsruhe. He was clearly obsessed by the problem and would spend long hours on train trips across Russia with a special deck of cards on which he had written each element and its atomic weight, playing what he called "chemical solitaire," shuffling, ordering, and reordering the elements. Yet when the solution finally came to him, it came during a time when he was not consciously trying to reach it.

memories seem to have little effect on the development and circuitry of the brain, how some traumatic memories remain perseveratively yet unchangingly active, while yet others become integrated and lead to profound and creative developments in the brain?

Creativity—that state when ideas seem to organize themselves into a swift, tightly woven flow, with a feeling of gorgeous clarity and meaning emerging—seems to me physiologically distinctive, and I think that if we had the ability to make fine enough brain images, these would show an unusual and widespread activity with innumerable connections and synchronizations occurring.

At such times, when I am writing, thoughts seem to organize themselves in spontaneous succession and to clothe themselves instantly in appropriate words. I feel I can bypass or transcend much of my own personality, my neuroses. It is at once not me and the innermost part of me, certainly the best part of me.

A General Feeling of Disorder

Nothing is more crucial to the survival and independence of organisms—be they elephants or protozoa—than the maintenance of a constant internal environment. Claude Bernard, the great French physiologist, said everything on this matter when, in the 1850s, he wrote, "La fixité du milieu intérieur est la condition de la vie libre." Maintaining such constancy is called homeostasis. The basics of homeostasis are relatively simple but miraculously efficient at the cellular level, where ion pumps in cell membranes allow the chemical interior of cells to remain constant, whatever the vicissitudes of the external environment. More complex monitoring systems are demanded when it comes to ensuring homeostasis in multicellular organisms—animals, and human beings, in particular.

Homeostatic regulation is accomplished by the development of special nerve cells and nerve nets (or plexuses) scattered throughout our bodies, as well as by direct chemical means (hormones, for example). These scattered nerve cells and plexuses become organized into a system or confederation that is largely autonomous in its functioning—hence its name, the autonomic nervous system. The autonomic nervous system was only recognized and explored in the early part of the twentieth century, whereas many of the functions of the central nervous system, especially the brain, had already been mapped in detail in the nineteenth century. This is something of a paradox, for the autonomic nervous system evolved long before the central nervous system.

They were (and to a considerable extent still are) independent evolutions, extremely different in organization as well as formation. Central nervous systems, along with muscles and sense organs, evolved to allow animals to get around in the world—forage, hunt, seek mates, avoid or fight enemies, etc. The central nervous system, along with the proprioceptive system, tells one who one is and what one is doing. The autonomic nervous system, sleeplessly monitoring every organ and tissue in the body, tells one *how* one is. (Curiously, the brain itself has no sense organs, which is why one can have gross disorders here yet feel no malaise. Thus Ralph Waldo Emerson, who developed Alzheimer's disease in his sixties, when asked how he was, would say, "I have lost my mental faculties but am perfectly well.")[1]

By the early twentieth century, two general divisions of the autonomic nervous system were recognized: a "sym-

1. David Shenk describes this beautifully in his book *The Forgetting.*

pathetic" part, which, by increasing the heart's output, sharpening the senses, and tensing the muscles, readies an animal for action (in extreme situations, for instance, life-saving fight or flight); and the corresponding opposite—a "parasympathetic" part—which increases activity in the "housekeeping" parts of the body (gut, kidneys, liver, etc.), slowing the heart and promoting relaxation and sleep. These two portions of the autonomic nervous system normally work in a happy reciprocity; thus the delicious postprandial somnolence that follows a heavy meal is not the state in which to run a race or get into a fight. When the two parts of the autonomic nervous system are working harmoniously together, one feels "well" or "normal."

No one has written more eloquently about this than Antonio Damasio in his book *The Feeling of What Happens* and many subsequent books and papers. He speaks of a "core consciousness," the basic feeling of *how one is,* which eventually becomes a dim, implicit feeling of consciousness.[2] It is especially when things are going wrong internally—when homeostasis is not being maintained; when the autonomic balance starts listing heavily to one side or the other—that this core consciousness, the feeling of *how one is,* takes on an intrusive, unpleasant quality, and now one will say, "I feel ill—something is amiss." At such times, one no longer *looks* well either.

2. See also Antonio Damasio and Gil B. Carvalho, "The Nature of Feelings: Evolutionary and Neurobiological Origins" (2013).

As an example of this, migraine is a sort of prototype illness, often very unpleasant but transient and self-limiting; benign in the sense that it does not cause death or serious injury and that it is not associated with any tissue damage or trauma or infection. Migraine provides, in miniature, the essential features of *being ill*—of trouble inside the body—without actual illness.

When I came to New York, nearly fifty years ago, the first patients I saw suffered from attacks of migraine—common migraine, so called because it attacks at least 10 percent of the population. (I myself have had attacks of them throughout my life.) Seeing such patients, trying to understand or help them, constituted my apprenticeship in medicine and led to my first book, *Migraine.*

Though there are many (one is tempted to say, innumerable) possible presentations of common migraine—I described nearly a hundred such in my book—its commonest harbinger may be just an indefinable but undeniable feeling of *something amiss.* This is exactly what Emil du Bois-Reymond emphasized when, in 1860, he described his own attacks of migraine. "I wake," he writes, "with a general feeling of disorder."

In his case (he had had migraines every three to four weeks since his twentieth year), there would be "a slight pain in the region of the right temple which . . . reaches its greatest intensity at midday; towards evening it usually passes off . . . At rest the pain is bearable, but it is increased by motion to a high degree of violence . . . It responds to each beat of the temporal artery." Moreover, du Bois-Reymond looked different during his migraines: "The countenance is pale and sunken, the right eye small and reddened." During violent attacks he would experience nausea and "gastric disorder." The "general feeling of disorder" that so often inaugurates migraines may continue, getting more and more severe in the course of an attack; the worst-affected patients may be reduced to lying in a leaden haze, feeling half-dead or even that death would be preferable.[3]

I cite du Bois-Reymond's self-description, as I do at the very beginning of *Migraine,* partly for its precision and beauty (as are common in nineteenth-century neurological descriptions, but rare now), but above all because it is *exemplary*—all cases of migraine vary, but they are, so to speak, permutations of his.

The vascular and visceral symptoms of migraine are typical of unbridled parasympathetic activity, but they may be preceded by a physiologically opposite state. One may feel full of energy, even a sort of euphoria, for a few hours *before* a migraine—George Eliot would speak of

3. Aretaeus noted in the second century that patients in such a state "are weary of life and wise to die." Such feelings, while they may originate, and be correlated, with autonomic imbalance, must connect with those "central" parts of the autonomic nervous system in which feeling, mood, sentience, and (core) consciousness are mediated—the brain stem, hypothalamus, amygdala, and other subcortical structures.

herself as feeling "dangerously well" at such times. There may, similarly, especially if the suffering has been very intense, be a "rebound" *after* a migraine. This was very clear with one of my patients (Case 68 in *Migraine*), a young mathematician with very severe migraines. For him the resolution of a migraine, accompanied by a huge passage of pale urine, was always followed by a burst of original mathematical thinking. "Curing" his migraines, we found, "cured" his mathematical creativity, and he elected, given this strange economy of body and mind, to keep both.

While this is the general pattern of a migraine, there can occur rapidly changing fluctuations and contradictory symptoms—a feeling that patients often call "unsettled." In this unsettled state (as I write in *Migraine*), "one may feel hot or cold, or both . . . bloated and tight, or loose and queasy; a peculiar tension, or languor, or both . . . sundry strains and discomforts, which come and go."

Indeed, everything comes and goes, and if one could take a scan or inner photograph of the body at such times, one would see vascular beds opening and closing, peristalsis accelerating or stopping, viscera squirming or tightening in spasms, secretions suddenly increasing or decreasing, as if the nervous system itself were in a state of indecision. Instability, fluctuation, and oscillation are of the essence in the unsettled state, this general feeling of disorder. We lose the normal feeling of "wellness," which all of us, and perhaps all animals, have in health.

If new thoughts about illness and recovery—or old thoughts in new form—have been stimulated by thinking back to my first patients, they have been given an unexpected salience by a very different personal experience in recent weeks.

On Monday, February 16, 2015, I could say I felt well, in my usual state of health—at least such health and energy as a fairly active eighty-one-year-old can hope to enjoy—and this despite learning, a month earlier, that much of my liver was occupied by metastatic cancer. Various palliative treatments had been suggested—treatments that might reduce the load of metastases in my liver and permit a few extra months of life. The one I opted for, decided to try first, involved my surgeon, an interventional radiologist, threading a catheter up to the bifurcation of the hepatic artery and then injecting a mass of tiny beads into the right hepatic artery, where they would be carried to the smallest arterioles, blocking these, cutting off the blood supply and oxygen needed by the metastases—in effect, starving and asphyxiating them to death. (My surgeon, who has a gift for vivid metaphor, compared this to killing rats in the basement, or, in a pleasanter image, mowing down the dandelions on the back lawn.) If such an embolization proved to be effective, and tolerated, it could be done on the other side of the liver (the dandelions on the front lawn) a month or so later.

The procedure, though relatively benign, would lead to the death of a huge mass of melanoma cells (almost 50 percent of my liver was occupied by metastases). These, in dying, would give off a variety of unpleasant and pain-producing substances and would then have to be removed, as all dead material must be removed from the body. This immense task of garbage disposal would be undertaken by cells of the immune system—macrophages—that are specialized to engulf alien or dead matter in the body. I might think of them, my surgeon suggested, as tiny spiders, millions or perhaps billions in number, scurrying inside me, engulfing the melanoma debris. This enormous cellular task would sap all my energy, and I would feel, in consequence, a tiredness beyond anything I had ever felt before, to say nothing of pain and other problems.

I am glad I was forewarned, for the following day (Tuesday, the seventeenth), soon after waking from the embolization—it was performed under general anesthesia—I was to be assailed by feelings of excruciating tiredness and paroxysms of sleep so abrupt they could poleax me in the middle of a sentence or a mouthful, or when visiting friends were talking or laughing loudly a yard away from me. Sometimes, too, delirium would seize me within seconds, even in the middle of handwriting. I felt extremely weak and inert; I would sometimes sit motionless until hoisted to my feet and walked by two helpers. While pain seemed tolerable at rest, an involuntary movement such as a sneeze or hiccup would produce an explosion, a sort of negative orgasm of pain, despite my being maintained, like all post-embolization patients, on a continuous intravenous infusion of narcotics. This massive infusion of narcotics halted all bowel activity for nearly a week, so that everything I ate—I had no appetite but had to "take nourishment," as the nursing staff put it—was retained inside me.

Another problem—not uncommon after the embolization of a large part of the liver—was a release of ADH, antidiuretic hormone, which caused an enormous accumulation of fluid in my body. My feet became so swollen they were almost unrecognizable *as* feet, and I developed a thick tire of edema around my trunk. This "hyperhydration" led to lowered levels of sodium in my blood, which probably contributed to my deliria. With all this, and a variety of other symptoms—temperature regulation was unstable, I would be hot one minute, cold the next—I felt awful. I had "a general feeling of disorder" raised to an almost infinite degree. If I had to feel like this from now on, I kept thinking, I would sooner be dead.

I stayed in the hospital for six days after embolization and then returned home. Although I still felt worse than I had ever felt in my life, I did in fact feel a little better, minimally better, with each passing day (and everyone told me, as they tend to tell sick people, that I was looking "great"). I still had sudden, overwhelming paroxysms of sleep, but I forced myself to work, correcting the galleys of my autobiography (even though I might fall asleep in mid-sentence, my head dropping heavily onto the desk, my hand still clutching a pen). These post-embolization days would have been very difficult to endure without this task (which was also a joy).

On day ten, I turned a corner—I felt awful, as usual, in the morning, but a completely different person in the afternoon. This was delightful, and wholly unexpected: there was no intimation, beforehand, that such a transformation was about to happen. I regained some appetite, my bowels started working again, and on February 28 and March 1, I had a huge and delicious diuresis, losing fifteen pounds over the course of two days. I suddenly found myself full of physical and creative energy and a euphoria almost akin to hypomania. I strode up and down the corridor in my apartment building while exuberant thoughts rushed through my mind.

How much of this was a reestablishment of balance in the body; how much an autonomic rebound after a profound autonomic depression; how much other physiological factors; and how much the sheer joy of writing, I do not know. But my transformed state and feeling were, I suspect, very close to what Nietzsche experienced after a period of illness and expressed so lyrically in *The Gay Science:*

> Gratitude pours forth continually, as if the unexpected had just happened—the gratitude of a convalescent—

for convalescence was unexpected. . . . The rejoicing of strength that is returning, of a reawakened faith in a tomorrow or the day after tomorrow, of a sudden sense and anticipation of a future, of impending adventures, of seas that are open again.

The River of Consciousness

"Time," says Jorge Luis Borges, "is the substance I am made of. Time is a river that carries me away, but I am the river." Our movements, our actions, are extended in time, as are our perceptions, our thoughts, the contents of consciousness. We live in time, we organize time, we are time creatures through and through. But is the time we live in, or live by, continuous, like Borges's river? Or is it more comparable to a succession of discrete moments, like beads on a string?

David Hume, in the eighteenth century, favored the idea of discrete moments, and for him the mind was "nothing but a bundle or collection of different perceptions, which succeed each other with an inconceivable rapidity, and are in a perpetual flux and movement."

For William James, writing his *Principles of Psychology* in 1890, the "Humean view," as he called it, was both powerful and vexing. It seemed counterintuitive, as a start. In his famous chapter on "the stream of thought," James stressed that to its possessor consciousness seems to be always continuous, "without breach, crack, or division," never "chopped up in bits." The content of consciousness might be changing continually, but we move smoothly from one thought to another, one percept to another, without interruption or breaks. For James, thought flowed, hence his introduction of the term "stream of consciousness." But, he wondered, "is consciousness really discontinuous . . . does it only seem continuous to itself by an illusion analogous to that of the zoetrope?"

Before about 1830 (short of making an actual working model), we had no way of making representations or images that had movement. Nor would it have occurred to most that a sensation or illusion of movement could be conveyed by still pictures. How could pictures convey movement if they had none themselves? The very idea was paradoxical, a contradiction. But the zoetrope proved that individual images could be fused in the brain to give an illusion of continuous motion.

Zoetropes (and many other similar devices with a variety of names) were extremely popular in James's time, and few middle-class Victorian households were without one. These instruments contained a drum or disc on which a sequence of drawings—"freeze frames" of animals moving, ball games, acrobats in motion, plants growing—was painted or pasted. When the drum or disc was rotated, the separate drawings flicked by in rapid succession and, at a critical speed, suddenly gave way to the perception of a single, steadily moving picture. Though zoetropes were popular as toys, providing a magical illusion of motion, they were originally designed (often by scientists or philosophers) to serve a very serious purpose: to illuminate the mechanisms of animal motion and of vision itself.

Had James been writing a few years later, he might have used the analogy of a motion picture. A movie, with its taut stream of thematically connected images, its visual narrative integrated by the viewpoint and values of its director, is not at all a bad metaphor for the stream of consciousness. The technical and conceptual devices of cinema—zooming, fading, dissolving, omission, allusion, association, and juxtaposition of all sorts—rather closely mimic the streamings and veerings of consciousness in many ways.

It is an analogy that Henri Bergson used in his 1907 book *Creative Evolution,* in which he devoted an entire section to "The Cinematographical Mechanism of Thought and the Mechanistic Illusion." But when Bergson spoke of "cinematography" as an elemental mechanism of brain and mind, it was, for him, a very special sort of cinematography, in that its "snapshots" were not isolable from one another but organically connected. In *Time and Free Will,* he wrote of such perceptual moments as "permeating one another," "melting into" one another, like the notes of a tune (as opposed to the "empty and succeeding beats of a metronome").

James too wrote of connectedness and articulation, and for him these moments are connected by the whole trajectory and theme of a life:

> The knowledge of some other part of the stream, past or future, near or remote, is always mixed in with our knowledge of the present thing.
>
> . . . These lingerings of old objects, these incomings of new, are the germs of memory and expectation, the retrospective and the prospective sense of time. They give that continuity to consciousness without which it could not be called a stream.

In the same chapter, on the perception of time, James quotes a fascinating speculation of James Mill (the father of John Stuart Mill) as to what consciousness might be like if it were discontinuous, a string of bead-like sensations and images, all separate:

> We never could have any knowledge except that of the present instant. The moment each of our sensations ceased it would be gone for ever, and we should be as if we had never been . . . we would be wholly incapable of acquiring experience.

James wonders whether existence would indeed be possible under these circumstances, with consciousness reduced to "a glow-worm spark . . . [with] all beyond in total darkness." This is precisely the condition of someone with amnesia, though the "moment" here may be a few seconds in length. When I described my amnesic patient Jimmie, the "Lost Mariner," in *The Man Who Mistook His Wife for a Hat,* I wrote,

> He is . . . isolated in a single moment of being, with a moat or lacuna of forgetting all round him. . . . He is a man without a past (or future), stuck in a constantly changing, meaningless moment.

Were James and Bergson intuiting a truth in comparing visual perception—and indeed, the flow of consciousness itself—to mechanisms like zoetropes and movie cameras? Does the eye/brain actually "take" perceptual stills and somehow fuse them to give a sense of continuity and motion? No clear answer was forthcoming during their lifetimes.

There is a rare but dramatic neurological disturbance that a number of my patients have experienced during attacks of migraine, when they may lose the sense of visual continuity and motion and see instead a flickering series of "stills." The stills may be clear-cut and sharp, succeeding one another without superimposition or overlap. But more commonly they are somewhat blurred, as with a too-long photographic exposure; they persist long enough that each is still visible when the next "frame" is seen, so that three or four frames, the earlier ones progressively fainter, are apt to be superimposed on each other. (This effect resembles some of Étienne-Jules Marey's "chronophotographs" of the 1880s, in which one sees a whole array of photographic moments or time frames superimposed on a single plate.)[1]

Such attacks are brief, rare, and not readily predicted or provoked, and perhaps for this reason I could find no good accounts of the phenomenon in the medical literature. When I wrote about them in my 1970 book, *Migraine,* I used the term "cinematographic vision" for them, for patients always compared them to films run too slow. I noted that the rate of flickering in these episodes seemed to be between six and twelve per second. There might also be, in cases of migraine delirium, a flickering of kaleidoscopic patterns or hallucinations. (The flickering might then accelerate to restore the appearance of normal motion.)

1. Étienne-Jules Marey, in France, like Eadweard Muybridge in the United States, pioneered the development of quick-fire, instantaneous serial photographs. While these could be arrayed around a zoetrope drum to provide a brief "movie," they could also be used to decompose movement, to investigate the temporal organization and biodynamics of animal and human motion. This was Marey's special interest as a physiologist, and for this purpose he preferred to superimpose his images—a dozen or twenty images, a second's worth—on a single plate. Such composite photographs, in effect, captured a span of time; this is why he called them "chronophotographs." Marey's photographs became the model for all subsequent scientific photographic studies of movement, and chronophotography was an inspiration to artists, too (one thinks of Duchamp's famous *Nude Descending a Staircase,* which Duchamp himself referred to as "a static image of movement").

Marta Braun surveys Marey's work in her fascinating monograph, *Picturing Time,* and Rebecca Solnit discusses Muybridge and his influences in *River of Shadows: Eadweard Muybridge and the Technological Wild West.*

This was a startling visual phenomenon for which, in the 1960s, there was no good physiological explanation. But I could not help wondering then whether visual perception might in a very real way be analogous to cinematography, taking in the visual environment in brief, instantaneous, static frames or "stills" and then, under normal conditions, fusing these to give visual awareness its usual movement and continuity—a fusion which, seemingly, was failing to occur in the very abnormal conditions of these migraine attacks.

Such visual effects may also occur in certain seizures, as well as in intoxications (especially with hallucinogens such as LSD). And there are other unusual visual effects that may occur. Moving objects may leave a trailing smear or wake, images may repeat themselves, and afterimages may be greatly prolonged.[2]

I heard similar accounts in the late 1960s from some of my postencephalitic patients when they were "awakened," and especially overexcited, by taking the drug L-dopa. Some patients described cinematic vision; others described extraordinary "standstills," sometimes hours

2. I have experienced this myself after drinking *sakau,* an intoxicant popular in Micronesia. I described some of the effects in a journal, and later in my book *The Island of the Colorblind:*

> Ghost petals ray out from a flower on our table, like a halo around it; when it is moved . . . it leaves a slight train, a visual smear . . . in its wake. Watching a palm waving, I see a succession of stills, like a film run too slow, its continuity no longer maintained.

long, in which visual flow was arrested—and even the stream of movement, of action, of thought itself.

These standstills were especially severe with Hester Y. Once I was called to the ward because Mrs. Y. had started a bath, and there was now water overflowing in the bathroom. I found her standing completely motionless in the middle of the flood.

She jumped when I touched her and asked, "What happened?"

"You tell me," I answered.

She said that she had started to run a bath for herself, and there was an inch of water in the tub . . . and then I touched her and she suddenly realized that the tub must have run over and caused a flood. She had been stuck, transfixed, at that perceptual moment when there was just an inch of water in the bath.

Such standstills showed that consciousness could be brought to a halt for substantial periods while automatic, nonconscious function—maintenance of posture or breathing, for example—continued as before.

Another striking example of perceptual standstill can be demonstrated with a common visual illusion, the Necker cube. Normally, when we look at this ambiguous perspective drawing of a cube, it switches perspective every few seconds, seeming first to project, then to recede, and no effort of will suffices to prevent this switching back and forth. The drawing itself does not change, nor does its retinal image. The switching is a purely cortical process, a conflict in consciousness itself, as it vacillates between two possible perceptual interpretations. This switching is seen in all normal subjects and can be observed with functional brain imaging. But a postencephalitic patient during a standstill state may see the same unchanging perspective for minutes or hours at a time.[3]

The normal flow of consciousness, it seemed, could not only be fragmented, broken into small, snapshot-like bits, but also suspended intermittently, for hours at a time. I found this even more puzzling and uncanny than cinematic vision, for it had been accepted almost axiomatically since the time of William James that consciousness, in its very nature, is ever changing and ever flowing. Now my own clinical experience had to cast doubt on even that.

Thus I was primed to be further fascinated when, in 1983, Josef Zihl and his colleagues in Munich published a single, very fully described case of motion blindness: a woman who, following a stroke, became permanently

3. As I explore in my book *Musicophilia*, music, with its rhythm and flow, can be of crucial importance in such freezings, allowing patients to resume their flow of movement, perception, and thought. Music sometimes seems to act as a sort of model or template for the sense of time and movement that such patients have temporarily lost. Thus a parkinsonian patient in the midst of a standstill may be enabled to move, even to dance, when music is played. Neurologists intuitively use musical terms here and speak of parkinsonism as a "kinetic stutter" and normal movement as "kinetic melody."

unable to perceive motion. (The stroke damaged highly specific areas of the visual cortex which physiologists have shown in experimental animals to be crucial for motion perception.) In this patient, whom they call L. M., there were "freeze frames" lasting several seconds, during which Mrs. M. would see a prolonged motionless image and be visually unaware of any movement around her, though her flow of thought and perception was otherwise normal. She might begin a conversation with a friend standing in front of her but not be able to see her friend's lips moving or facial expressions changing. And if the friend moved around behind her, Mrs. M. might continue to "see" him in front of her, even though his voice now came from behind. She might see a car "frozen" a considerable distance away but find, when she tried to cross the road, that it was now almost upon her. She would see a "glacier," a frozen arc of tea coming from the spout of the teapot, but then realize that she had overfilled the cup and there was now a puddle of tea on the table. Such a condition was utterly bewildering and sometimes quite dangerous.

There are clear differences between cinematic vision and the sort of motion blindness described by Zihl, and perhaps between these and the very long visual and sometimes global freezes experienced by some postencephalitic patients. These differences imply that there must be a number of different mechanisms or systems for the perception of visual motion and the continuity of visual consciousness, and this accords with evidence obtained from perceptual and psychological experiments. Some or all of these mechanisms may fail to work as they should in certain intoxications, some attacks of migraine, and some forms of brain damage—but can they also reveal themselves under normal conditions?

An obvious example springs to mind, which many of us have seen and perhaps puzzled over when watching evenly rotating objects—fans, wheels, propeller blades—or when walking past fences or palings, when the normal continuity of motion seems to be interrupted. Thus occasionally as I lie in bed looking up at my ceiling fan, the blades seem suddenly to reverse direction for a few seconds and then to return equally suddenly to their original forward motion. Sometimes the fan seems to hover or stall, and sometimes to develop additional blades or dark bands broader than the blades.

It is similar to what happens in a film when the wheels of stagecoaches sometimes appear to be going slowly backwards or scarcely moving. This wagon-wheel illusion, as it is called, reflects a lack of synchronization between the rate of filming and that of the rotating wheels. But I can have a real-life wagon-wheel illusion when I look at my fan with the morning sun flooding into my room, bathing everything in a continuous, even light. Is there, then, some flickering or lack of synchronization in my own perceptual mechanisms—analogous, again, to the action of a movie camera?

Dale Purves and his colleagues have explored wagon-wheel illusions in great detail, and they have confirmed that this type of illusion or misperception is universal among their subjects. Having excluded any other cause of discontinuity (intermittent lighting, eye movements, etc.), they conclude that the visual system processes information "in sequential episodes," at the rate of three to twenty such episodes per second. Normally, these sequential images are experienced as an unbroken perceptual flow. Indeed, Purves suggests, we may find movies convincing precisely because we ourselves break up time and reality much as a movie camera does, into discrete frames, which we then reassemble into an apparently continuous flow.

In Purves's view, it is precisely this decomposition of what we see into a succession of moments that enables the brain to detect and compute motion, for all it has to do is to note the differing positions of objects between successive "frames" and from these calculate the direction and speed of motion.

~~~~

But this is not enough. We do not merely calculate movement as a robot might; we *perceive* it. We perceive motion, just as we perceive color or depth, as a
~~~~

unique qualitative experience that is vital to our visual awareness and consciousness. Something beyond our understanding occurs in the genesis of qualia, the transformation of an objective cerebral computation to a subjective experience. Philosophers argue endlessly over how these transformations occur and whether we will ever be capable of understanding them.

James imagined a zoetrope as a metaphor for the conscious brain, and Bergson compared it to cinematography, but these were, of necessity, no more than tantalizing analogies and images. It has only been in the last twenty or thirty years that neuroscience could even start to address such issues as the neural basis of consciousness.

From having been an almost untouchable subject before the 1970s, the neuroscientific study of consciousness has now become a central concern, one that engages scientists all over the world. Every level of consciousness is now being explored, from the most elemental perceptual mechanisms (mechanisms common to many animals besides ourselves) to the higher reaches of memory, imagery, and self-reflective consciousness.

Is it possible to define the almost inconceivably complex processes that form the neural correlates of thought and consciousness? We must imagine, if we can, that in our brains with their hundred billion neurons, each with a thousand or more synaptic connections, there may emerge or be selected, within fractions of a second, a million-odd neuronal groups or coalitions, each with a thousand or ten thousand neurons apiece. (Edelman speaks here of the "hyperastronomical" magnitudes involved.) All of these coalitions, like the "millions of flashing shuttles" in Sherrington's enchanted loom, are in communication with each other, weaving, many times a second, their continuously changing but always meaningful patterns.

We cannot begin to catch the density, the multifariousness of it all, the superimposed and mutually influencing layers of the stream of consciousness as it flows, constantly changing, through the mind. Even the highest powers of art—whether in film, or theater, or literary narrative—can only convey the faintest intimation of what human consciousness is really like.

It is now possible to monitor simultaneously the activities of a hundred or more individual neurons in the brain and to do so in unanesthetized animals given simple perceptual and mental tasks. We can examine the activity and interactions of large areas of the brain by means of imaging techniques like functional MRIs and PET scans, and such noninvasive techniques can be used with human subjects to see which areas of the brain are activated in complex mental activities.

In addition to physiological studies, there is the relatively new realm of computerized neural modeling, using populations or networks of virtual neurons and seeing how these organize themselves in response to various stimuli and constraints.

All of these approaches, along with concepts not available to earlier generations, now combine to make the quest for the neural correlates of consciousness the most fundamental and exciting adventure in neuroscience today. A crucial innovation has been population thinking, thinking in terms that take account of the brain's huge population of neurons and the power of experience to differentially alter the strengths of connections between them, and to promote the formation of functional groups or constellations of neurons throughout the brain—groups whose interactions serve to categorize experience.

Instead of seeing the brain as rigid, fixed in mode, programmed like a computer, there is now a much more biological and powerful notion of "experiential selection," of experience literally shaping the connectivity and function of the brain (within genetic, anatomical, and physiological limits).

Such a selection of neuronal groups (groups consisting of perhaps a thousand or so individual neurons) and its effect on shaping the brain over the lifetime of an individual, is seen as analogous to the role of natural selection in the evolution of species; hence Gerald M. Edelman, who was a pioneer in such thinking in the 1970s, speaks of "neural Darwinism." Jean-Pierre Changeux, more concerned with the connections of individual neurons, speaks of "the Darwinism of synapses."

William James himself always insisted that consciousness was not a "thing" but a "process." The neural basis of these processes, for Edelman, is one of dynamic interaction between neuronal groups in different areas of the cortex, as well as between the cortex and the thalamus and other parts of the brain. Edelman sees consciousness as arising from the enormous number of reciprocal interactions between memory systems in the anterior parts of the brain and systems concerned with perceptual categorization in the posterior parts of the brain.[4]

Francis Crick and his colleague Christof Koch have also been pioneers in the study of the neural basis of

4. No paradigms or concepts, however original, ever come totally out of the blue. While population thinking in relation to the brain only emerged in the 1970s, there was an important antecedent twenty-five years earlier: Donald Hebb's famous 1949 book, *The Organization of Behavior.* Hebb sought to bridge the great gap between neurophysiology and psychology with a general theory that could relate neural processes to mental ones and, in particular, show how experience could modify the brain. The potential for modification, Hebb felt, was vested in the synapses that connect brain cells to each other. Hebb's original concept was soon to be confirmed and set the stage for new ways of thinking. A single cerebral neuron, we now know, can have up to ten thousand synapses, and the brain as a whole has upwards of a hundred trillion, so the capacities for modification are practically infinite. Every neuroscientist who now thinks about consciousness is thus indebted to Hebb.

consciousness. From their first collaborative work in the 1980s, they have focused more narrowly on elementary visual perception and processes, feeling that the visual brain was the most amenable to investigation and could serve as a model for investigating and understanding higher and higher forms of consciousness.[5]

In a synoptic 2003 paper called "A Framework for Consciousness," Crick and Koch speculated on the neural correlates of motion perception, how visual continuity is perceived or constructed, and, by extension, the seeming continuity of consciousness itself. They proposed that "conscious awareness [for vision] is a series of static snapshots, with motion 'painted' on them . . . [and] that perception occurs in discrete epochs."

I was startled when I first came across this passage, because their formulation seemed to rest upon the same notion of consciousness that James and Bergson had intimated a century before, and the same notion that had been in my mind since I first heard accounts of cinematic vision from my migraine patients in the 1960s. Here, however, was something more, a possible substrate for consciousness based in neuronal activity.

The "snapshots" that Crick and Koch postulate are not uniform, like cinematic ones. The duration of suc-

5. Koch gives a vivid and personal history of their work, and of the search for the neural basis of consciousness generally, in his book *The Quest for Consciousness.*

cessive snapshots, they feel, is not likely to be constant; moreover, the time of a snapshot for shape, say, may not coincide with one for color. While this "snapshotting" mechanism for visual sensory inputs is probably a fairly simple and automatic one, a relatively low-order neural mechanism, each percept must include a great number of visual attributes, all of which are bound together on some preconscious level.[6]

How, then, are the various snapshots "assembled" to achieve apparent continuity, and how do they reach the level of consciousness?

While the perception of a particular motion (for example) may be represented by neurons firing at a particular rate in the motion centers of the visual cortex, this is only the beginning of an elaborate process. To reach consciousness, this neuronal firing, or some higher representation of it, must cross a certain threshold of intensity and be maintained above it; consciousness, for Crick and Koch, is a threshold phenomenon. To do that, this group of neurons must engage other parts of the brain (usually in the frontal lobes) and ally itself with millions of other

6. One hypothesis to explain the mechanisms of binding entails the synchronization of neuronal firing in a range of sensory areas. Sometimes it may fail to occur, and Crick cites a comic instance of this in his 1994 book, *The Astonishing Hypothesis:* "A friend walking in a busy street 'saw' a colleague and was about to address him when he realized that the black beard belonged to another passerby and the bald head and spectacles to another."

neurons to form a "coalition." Such coalitions, they conceive, can form and dissolve in a fraction of a second and involve reciprocal connections between the visual cortex and many other areas of the brain. These neural coalitions in different parts of the brain talk to one another in a continuous back-and-forth interaction. A single conscious visual percept may thus entail the parallel and mutually influencing activities of billions of nerve cells.

Finally, the activity of a coalition, or coalition of coalitions, if it is to reach consciousness, must not only cross a threshold of intensity but also be held there for a certain time—roughly a hundred milliseconds. This is the duration of a "perceptual moment."[7]

To explain the apparent continuity of visual consciousness, Crick and Koch suggest that the activity of the coalition shows "hysteresis," that is, a persistence outlasting the stimulus. This notion is very similar, in a way, to the "persistence of vision" theories advanced in the nineteenth century.[8] In his *Treatise on Physiological*

7. The term "perceptual moment" was first used by the psychologist J. M. Stroud in the 1950s, in his paper "The Fine Structure of Psychological Time." The perceptual moment represented for him the "grain" of psychological time, that duration (about a tenth of a second, he estimated from his experiments) which it took to integrate sensory information as a unit. But, as Crick and Koch remark, Stroud's "perceptual moment" hypothesis was virtually ignored for the next half a century.

8. In his delightful book *A Natural History of Vision,* Nicholas Wade quotes Seneca, Ptolemy, and other classical authors who,

Optics of 1860, Hermann von Helmholtz wrote, "All that is necessary is that the repetition of the impression shall be fast enough for the after-effect of one impression not to have died down perceptibly before the next one comes." Helmholtz and his contemporaries supposed that this aftereffect occurred in the retina, but for Crick and Koch it occurs in the coalitions of neurons in the cortex. The sense of continuity, in other words, results from the continuous overlapping of successive perceptual moments. It may be that the forms of cinematographic vision I have described—with either sharply separated stills or blurred and overlapping ones—represent abnormalities of excitability in the coalitions, with either too much or too little hysteresis.[9]

Vision, in ordinary circumstances, is seamless and gives no indication of the underlying processes on which it depends. It has to be decomposed, experimentally or in

---

observing that a flaming torch swung rapidly in a circle appeared to form a continuous ring of fire, realized that there must be a considerable duration or persistence of visual images (or, in Seneca's term, a "slowness" of vision). An impressively accurate measurement of this duration—as 8⁄60 of a second—was made in 1765, but it was only in the nineteenth century that the persistence of vision was systematically exploited in such instruments as the zoetrope. It seems too that motion illusions akin to the wagon-wheel effect were well known as much as two thousand years ago.

9. An alternative explanation, Crick and Koch suggest (personal communication), is that the blurring and persistence of snapshots is due to their reaching short-term memory (or a short-term visual memory buffer) and slowly decaying there.

neurological disorders, to show the elements that compose it. The flickering, perseverative, time-blurred images experienced in certain intoxications or severe migraines lend credence to the idea that consciousness is composed of discrete moments.

Whatever the mechanism, the fusing of discrete visual frames or snapshots is a prerequisite for continuity, for a flowing, mobile consciousness. Such a dynamic consciousness probably first arose in reptiles a quarter of a billion years ago. It seems probable that no such stream of consciousness exists in amphibians. A frog, for example, shows no active attention and no visual following of events. The frog does not have a visual world or visual consciousness as we know it, only a purely automatic ability to recognize an insect-like object if one enters its visual field and to dart out its tongue in response. It does not scan its surroundings or look for prey.

If a dynamic, flowing consciousness allows, at the lowest level, a continuous active scanning or looking, at a higher level it allows the interaction of perception and memory, of present and past. And such a "primary" consciousness, as Edelman calls it, is highly efficacious, highly adaptive, in the struggle for life.

In his book *Wider Than the Sky: The Phenomenal Gift of Consciousness,* Edelman writes,

> Imagine an animal with primary consciousness in the jungle. It hears a low growling noise, and at

> the same time the wind shifts and the light begins to wane. It quickly runs away, to a safer location. A physicist might not be able to detect any necessary causal relation among these events. But to an animal with primary consciousness, just such a set of simultaneous events might have accompanied a previous experience, which included the appearance of a tiger. Consciousness allowed integration of the present scene with the animal's past history of conscious experience, and that integration has survival value whether a tiger is present or not.

From such a relatively simple primary consciousness, we leap to human consciousness, with the advent of language and self-consciousness and an explicit sense of the past and the future. And it is this which gives a thematic and personal continuity to the consciousness of every individual. As I write, I am sitting at a café on Seventh Avenue, watching the world go by. My attention and focus dart to and fro: a girl in a red dress goes by, a man walking a funny dog, the sun (at last!) emerging from the clouds. But there are also other sensations that seem to come by themselves: the noise of a car backfiring, the smell of cigarette smoke as an upwind neighbor lights up. These are all events which catch my attention for a moment as they happen. Why, out of a thousand possible perceptions, are these the ones I seize upon? Reflections, memories, associations, lie behind them. For consciousness is always active and selective—charged with feelings and meanings uniquely our own, informing our choices and interfusing our perceptions. So it is not just Seventh Avenue that I see but *my* Seventh Avenue, marked by my own selfhood and identity.

Christopher Isherwood starts "A Berlin Diary" with an extended photographic metaphor: "I am a camera with its shutter open, quite passive, recording, not thinking. Recording the man shaving at the window opposite and the woman in the kimono washing her hair. Some day, all this will have to be developed, carefully printed, fixed." But we deceive ourselves if we imagine that we can ever be passive, impartial observers. Every perception, every scene, is shaped by us, whether we intend it or know it, or not. We are the directors of the film we are making—but we are its subjects too: every frame, every moment, is us, is ours.

But how then do our frames, our momentary moments, hold together? How, if there is only transience, do we achieve continuity? Our passing thoughts, as William James says (in an image that smacks of cowboy life in the 1880s), do not wander round like wild cattle. Each one is owned and bears the brand of this ownership, and each thought, in James's words, is born an owner of the thoughts that went before, and "dies owned, transmitting whatever it realized as its Self to its own later proprietor."

So it is not just perceptual moments, simple physiological moments—though these underlie everything else—but moments of an essentially personal kind that seem to constitute our very being. Finally, then, we come around to Proust's image, itself slightly reminiscent of photography, that we consist entirely of "a collection of moments," even though these flow into one another like Borges's river.

Scotoma: Forgetting and Neglect in Science

We may look at the history of ideas backwards or forwards: we can retrace the earlier stages, the intimations, and the anticipations of what we think now; or we can concentrate on the evolution, the effects and influences of what we once thought. Either way, we may imagine that history will be revealed as a continuum, an advance, an opening like Darwin's tree of life. What one often finds, however, is very far from a majestic unfolding, and very far from being a continuum in any sense.

I began to realize how elusive scientific history can be when I became involved with my first love, chemistry. I vividly remember, as a boy, reading a history of chemistry and learning that what we now call oxygen had been all but discovered in the 1670s by John Mayow, a century before Scheele and Priestley identified it. Mayow, through careful experimentation, showed that approximately one-fifth of the air we breathe consists of a substance necessary to both combustion and respiration (he called it "spiritus nitro-aereus"). And yet Mayow's prescient work, widely read in his time, was somehow forgotten and obscured by the competing phlogiston theory, which prevailed for another century until Lavoisier finally disproved it in the 1780s. Mayow had died a hundred years earlier, at the age of thirty-nine. "Had he lived but a little longer," the author of this history, F. P. Armitage, wrote, "it can scarcely be doubted that he would have forestalled the revolutionary work of Lavoisier, and stifled the theory of phlogiston at its birth." Was this a romantic exaltation of John Mayow, a romantic misreading of the structure of the scientific enterprise, or could the history of chemistry have been wholly different, as Armitage suggested?[1]

Such forgetting or neglect of history is not uncommon

1. Armitage, a former master at my own school, published his book in 1906 to stimulate the enthusiasm of Edwardian schoolboys, and it seems to me now, with different eyes, that it has a somewhat romantic and jingoistic ring, an insistence that it was the English, not the French, who discovered oxygen.

William Brock, in his *History of Chemistry,* provides a different perspective. "Early historians of chemistry liked to find a close resemblance between Mayow's explanation and the later oxygen theory of calcination," he writes. But such resemblances, Brock stresses, "are superficial, for Mayow's theory was a mechanical, not a chemical, theory of combustion. . . . It marked a return to a dualistic world of principles and occult powers."

All the greatest innovators of the seventeenth century, not excluding Newton, still had one foot in the medieval world of alchemy, the hermetic, and the occult—indeed Newton's intense interest in alchemy and the esoteric continued to the end of his life. (This fact was largely forgotten until John Maynard Keynes brought it out, startlingly, in his 1946 essay "Newton, the Man," but the overlap of "modern" and "occult" in the climate of seventeenth-century science is now well accepted.)

in science; I saw it for myself when I was a young neurologist just starting work in a headache clinic. My job was to make a diagnosis—migraine, tension headache, whatever—and prescribe a treatment. But I could never confine myself to this, nor could many of the patients I saw. They would often tell me, or I would observe, other phenomena: sometimes distressing, sometimes intriguing, but not strictly part of the medical picture—not needed, at least to make a diagnosis.

Often a classical visual migraine is preceded by an aura, so called, where the patient may see brightly scintillating zigzags slowly traversing the field of vision. These are well described and understood. But more rarely, patients would tell me of complex geometrical patterns that appeared in place of, or in addition to, the zigzags: lattices, whorls, funnels, and webs, constantly shifting, gyrating, and modulating. When I searched the current literature, I could find no mention of these. Puzzled, I decided to look at nineteenth-century accounts, which tend to be much fuller, much more vivid, much richer in description, than modern ones.

My first discovery was in the rare-book section of our college library (everything written before 1900 counted as "rare")—an extraordinary book on migraine written in the 1860s by a Victorian physician, Edward Liveing. It had a wonderful, lengthy title, *On Megrim, Sick-Headache, and Some Allied Disorders: A Contribution to the Pathology of Nerve-Storms,* and it was a grand, meandering sort of book, clearly written in an age far more leisurely, less rigidly constrained, than ours. It touched briefly on the complex geometrical patterns many of my patients had described, and it referred me to an 1858 paper, "On Sensorial Vision," by John Frederick Herschel, an eminent astronomer. I felt I had struck pay dirt at last. Herschel gave meticulous, elaborate descriptions of exactly the phenomena my patients had described; he had experienced them himself, and he ventured some deep speculations about their possible nature and origin. He thought they might represent "a sort of kaleidoscopic power" in the sensorium, a primitive, pre-personal generating power in the mind, the earliest stages, even precursors, of perception.

I could find no adequate description of these "geometrical spectra," as Herschel called them, in the entire century between his observations and my own—and yet it was clear to me that perhaps one person in twenty affected with visual migraine experienced them on occasion. How had these phenomena—startling, highly characteristic, unmistakable hallucinatory patterns—evaded notice for so long?

In the first place, someone must make an observation and report it. In 1858, the same year that Herschel reported his "spectra," Guillaume Duchenne, a French neurologist, published a detailed description of a boy with what we now call muscular dystrophy, followed a year later by a report on thirteen more cases. His observations rapidly entered the mainstream of clinical neurology, identified as a disorder of great importance. Physicians started "seeing" the dystrophy everywhere, and within a few years scores of further cases had been published in the medical literature. The disorder had always existed, ubiquitous and unmistakable, but very few physicians had reported on it before Duchenne.[2]

Herschel's paper on hallucinatory patterns, by contrast, sank without a trace. Perhaps this was because he was not a physician making medical observations but simply an independent observer of great curiosity. Though he suspected that his observations had scientific importance—that such phenomena could lead to deep insights about the brain—their medical importance was not his focus. His paper was published not in a medical journal but in a general scientific one. Because migraine was usually defined as a "medical" condition, Herschel's descriptions were not seen as relevant, and after a brief mention in Liveing's book they were forgotten or ignored by the medical profession. In a sense, Herschel's observations were premature; if they were to point to new scientific ideas about the mind and brain, there was no way of making such a connection in the 1850s—the necessary

2. Duchenne's most famous student, Jean-Martin Charcot, remarked,

> How is it that a disease so common, so widespread, and so recognizable at a glance . . . is recognized only now? Why did we need M. Duchenne to open our eyes?

concepts only emerged more than a century later with the development of chaos theory in the 1970s and 1980s.

According to chaos theory, although it is impossible to predict the individual behavior of each element in a complex dynamic system (for instance, the individual neurons or neuronal groups in the primary visual cortex), patterns can be discerned at a higher level by using mathematical models and computer analyses. There are "universal behaviors" which represent the ways such dynamic, nonlinear systems self-organize. These tend to take the form of complex reiterative patterns in space and time—indeed the very sorts of networks, whorls, spirals, and webs that one sees in the geometrical hallucinations of migraine.

Such chaotic, self-organizing behaviors have now been recognized in a vast range of natural systems, from the eccentric motions of Pluto to the striking patterns that appear in the course of certain chemical reactions to the multiplication of slime molds or the vagaries of weather. With this, a hitherto insignificant or unregarded phenomenon like the geometrical patterns of migraine aura suddenly assumes a new importance. It shows us, in the form of a hallucinatory display, not only an elemental activity of the cerebral cortex but an entire self-organizing system, a universal behavior, at work.[3]

3. When I described the phenomena of migraine aura in the original 1970 edition of my book *Migraine,* I could only say that they were

With migraine, I had to go back to an earlier, forgotten medical literature—a literature that most of my colleagues saw as superseded or obsolete. I found myself in a similar position with Tourette's. My interest in this syndrome had been kindled in 1969 when I was able to "awaken" a number of postencephalitic patients with L-dopa and saw how many of them rapidly swung from motionless, trancelike states through a tantalizing brief "normality" and then to the opposite extreme—violently hyperkinetic, tic-ridden states very similar to the half-mythical "Tourette's syndrome." I say "half-mythical" because no one in the 1960s spoke much about Tourette's; it was considered extremely rare and possibly factitious. I had only vaguely heard of it.

Indeed, in 1969, when I started to think about it, as my own patients were becoming palpably tourettic, I had difficulty finding any current references, and once again had to go back to the literature of the previous century: to Gilles de la Tourette's original papers in 1885 and 1886 and to the dozen or so reports that followed them. It was an era of superb, mostly French, descriptions of the varie-

"inexplicable" by existing concepts. But by 1992, in a revised edition and with the help of my colleague Ralph M. Siegel, I was able to add a chapter discussing these phenomena in the new light of chaos theory.

ties of tic behavior, which culminated in the book *Les tics et leur traitement* published in 1902 by Henri Meige and E. Feindel. Yet between 1907, when their book was translated into English, and 1970, the syndrome itself seemed almost to have disappeared.

Why? One must wonder whether this neglect was not caused by the growing pressures at the beginning of the new century to try to explain scientific phenomena, following a time when it was enough to simply describe them. And Tourette's was peculiarly difficult to explain. In its most complex forms it could express itself not only as convulsive movements and noises but as tics, compulsions, obsessions, and tendencies to make jokes and puns, to play with boundaries, and engage in social provocations and elaborate fantasies. Though there were attempts to explain the syndrome in psychoanalytical terms, these, while casting light on some of the phenomena, were impotent to explain others; there were clearly organic components as well. In 1960, the finding that a drug, haloperidol, which counters the effects of dopamine, could extinguish many of the phenomena of Tourette's generated a much more tractable hypothesis—that Tourette's was essentially a chemical disease, caused by an excess of (or an excessive sensitivity to) the neurotransmitter dopamine.

With this comfortable, reductive explanation to hand, the syndrome suddenly sprang into prominence again and indeed seemed to multiply its incidence a thousandfold. (It is currently considered to affect one person in a hundred.) There is now a very intensive investigation of Tourette's syndrome, but it is an investigation largely confined to molecular and genetic aspects. And while these may explain some of the overall excitability of Tourette's, they do little to illuminate the particular forms of the tourettic disposition to engage in comedy, fantasy, mimicry, mockery, dream, exhibition, provocation, and play. While we have moved from an era of pure description to one of active investigation and explanation, Tourette's itself has been fragmented in the process and is no longer seen as a whole.

This sort of fragmentation is perhaps typical of a certain stage in science—the stage that follows pure description. But the fragments must somehow, sometime, be gathered together and presented once more as a coherent whole. This requires an understanding of determinants at every level, from the neurophysiological to the psychological to the sociological—and of their continuous and intricate interaction.[4]

4. A somewhat similar sequence has occurred in "medical" psychiatry. If one looks at the charts of patients institutionalized in asylums and state hospitals in the 1920s and 1930s, one finds extremely detailed clinical and phenomenological observations, often embedded in narratives of an almost novelistic richness and density (as in the classical descriptions of Kraepelin and others at the turn of the century). With the institution of rigid diagnostic criteria and manuals (the *Diagnostic and Statistical Manuals*, or *DSMs*) this richness and detail and phenomenological openness have disappeared, and one

~~~~

In 1974, after I had spent fifteen years as a physician making observations on patients' neurological conditions, I had a neuropsychological experience of my own. I had severely injured the nerves and muscles of my left leg while climbing in a remote part of Norway; I needed surgery to repair the muscle tendons and time to allow the healing of nerves. During the two-week period after surgery, while my leg was immobilized in a cast, bereft of movement and sensation, it ceased to feel like a part of me. It seemed to have become a lifeless object, not real, not mine, inconceivably alien. But when I tried to communicate this feeling to my surgeon, he said, "Sacks, you're unique. I've never heard of anything like this from a patient before."

I found this absurd. How could I be "unique"? There must be other cases, I thought, even if my surgeon had not heard of them. As soon as I was mobile enough, I started to talk to my fellow patients, and many of them, I found,

---

finds instead meager notes that give no real picture of the patient or his world but reduce him and his disease to a list of "major" and "minor" diagnostic criteria. Present-day psychiatric charts in hospitals are almost completely devoid of the depth and density of information one finds in the older charts and will be of little use in helping us to bring about the synthesis of neuroscience with psychiatric knowledge that we so need. The "old" case histories and charts, however, will remain invaluable.
~~~~

had similar experiences of "alien" limbs. Some had found this so uncanny and fearful that they had tried to put it out of their minds; others had worried about it secretly but not tried to describe it to others.

After I left the hospital, I went to the library, determined to seek out the literature on the subject. For three years I found nothing. Then I came across an account by Silas Weir Mitchell, an American neurologist working at a Philadelphia hospital for amputees during the Civil War. He described, very fully and carefully, the phantom limbs (or "sensory ghosts," as he called them) that amputees experienced in place of their lost limbs. He also wrote of "negative phantoms," the subjective annihilation and alienation of limbs following severe injury and surgery. He was so struck by these phenomena that he wrote a special circular on the matter, which was distributed by the surgeon general's office in 1864.

Weir Mitchell's observations aroused brief interest but then disappeared. More than fifty years passed before the syndrome was rediscovered as thousands of new cases of neurological trauma were treated during the First World War. In 1917, the French neurologist Joseph Babinski (with Jules Froment) published a monograph in which, apparently ignorant of Weir Mitchell's report, he described the syndrome I had experienced with my own leg injury. Babinski's observations, like Weir Mitchell's, sank without a trace. (When, in 1975, I finally came upon Babinski's book in our library, I found I was the first person to have borrowed it since 1918.) During the Second World War, the syndrome was fully and richly described for a third time by two Soviet neurologists, Aleksei N. Leont'ev and Alexander Zaporozhets—again in ignorance of their predecessors. Yet though their book, *Rehabilitation of Hand Function*, was translated into English in 1960, their observations completely failed to enter the consciousness of either neurologists or rehabilitation specialists.[5]

The work of Weir Mitchell and Babinski, of Leont'ev and Zaporozhets, seemed to have fallen into a historical or cultural scotoma, a "memory hole," as Orwell would say.

As I pieced together this extraordinary, even bizarre story, I felt more sympathy with my surgeon's saying that he had never heard of anything like my symptoms before. The syndrome is not that uncommon: it occurs whenever there is a significant loss of proprioception and other sensory feedback through immobility or nerve damage. But why is it so difficult to put this on record, to give the syndrome its due place in our neurological knowledge and consciousness?

As used by neurologists, the term "scotoma" (from the Greek for "darkness") denotes a disconnection or hiatus in perception, essentially a gap in consciousness produced by a neurological lesion. (Such lesions may be

5. The study and understanding of phantom limbs has been given fresh impetus over the last few decades by the very large number of wartime amputees, fueling more research and the burgeoning technology of modern prostheses. I describe phantom limb syndrome in greater detail in my book *Hallucinations*.

at any level, from the peripheral nerves, as in my own case, to the sensory cortex of the brain.) It is extremely difficult for a patient with such a scotoma to be able to communicate what is happening. He himself scotomizes the experience because the affected limb is no longer part of his internal body image. Such a scotoma is literally unimaginable unless one is actually experiencing it. This is why I suggest, only half jocularly, that people read *A Leg to Stand On* while under spinal anesthesia, so that they will know in their own persons what I am describing.

Let us turn from this uncanny realm of alien limbs to a more positive phenomenon (but still a strangely neglected and scotomized one)—that of acquired cerebral achromatopsia or total color blindness following a cerebral injury or lesion. (This is a completely different condition from common color blindness, which is caused by a deficiency of one or more color receptors in the retina.) I choose this example because I have explored it in some detail, after I learned of it quite by accident, when a patient with the condition wrote to me.[6]

6. Mr. I., a painter, had normal color vision until he had a car accident and suddenly lost all sense of color—thus he had "acquired"

When I looked into the history of achromatopsia, I again encountered a remarkable gap or anachronism. Acquired cerebral achromatopsia—and even more dramatically, hemi-achromatopsia, the loss of color perception in only one half of the visual field, coming on suddenly as a consequence of a stroke—had been described in exemplary fashion in 1888 by a Swiss neurologist, Louis Verrey. When his patient subsequently died and came to autopsy, Verrey was able to delineate the exact area of the visual cortex that had been damaged by her stroke. Here, he predicted, "the center for chromatic sense will be found." Within a few years of Verrey's report, there were other careful reports of similar problems with color perception and the lesions that caused them. Achromatopsia and its neural basis seemed firmly established. But then, strangely, the literature fell silent—not a single full case report was published for another seventy-five years.

This story has been discussed with great scholarship and acumen by both Antonio Damasio and Semir Zeki.[7] Zeki remarks that Verrey's findings aroused resistance the moment they were published and sees their virtual

---

achromatopsia, as I describe in *An Anthropologist on Mars*. But there are also people who are congenitally achromatopic, as I explore in *The Island of the Colorblind*.

7. For Damasio's appraisal, see his 1980 paper in *Neurology*, "Central Achromatopsia: Behavioral, Anatomic, and Physiologic Aspects." Zeki's history of Verrey and others appears in a 1990 review paper in *Brain*, "A Century of Cerebral Achromatopsia."

denial and dismissal as springing from a deep and perhaps unconscious philosophical attitude—the then prevailing belief in the seamlessness of vision.

The notion that we are given the visual world as a datum, an image, complete with color, form, movement, and depth, is a natural and intuitive one, seemingly supported by Newtonian optics and Lockean sensationalism. The invention of the camera lucida, and later of photography, seemed to exemplify such a mechanical model of perception. Why should the brain behave any differently? Color, it was obvious, was an integral part of the visual image and not to be dissociated from it. The ideas of an isolated loss of color perception or of a center for chromatic sensation in the brain were thought self-evident nonsense. Verrey had to be wrong; such absurd notions had to be dismissed out of hand. So they were, and achromatopsia "disappeared."

There were, of course, other factors at work as well. Damasio has described how, in 1919, when Gordon Holmes published his findings on two hundred cases of war injuries to the visual cortex, he stated summarily that none of these showed isolated deficiencies in color perception. Holmes was a man of formidable authority and power in the neurological world, and his empirically based antagonism to the notion of a color center in the brain, reiterated with increasing force for over thirty years, was a major factor in preventing other neurologists from recognizing the syndrome.

The notion of perception as "given" in some seamless, overall way was finally shaken to its foundations in the late 1950s and early 1960s when David Hubel and Torsten Wiesel showed that there were cells and columns of cells in the visual cortex which acted as "feature detectors," specifically sensitive to horizontals, verticals, edges, alignments, or other features of the visual field. The idea began to develop that vision had components, that visual representations were in no sense "given," like optical images or photographs, but were constructed by an enormously complex and intricate correlation of different processes. Perception was now seen as composite, as modular, the interaction of a huge number of components. The integration and seamlessness of perception had to be achieved in the brain.

It thus became clear in the 1960s that vision was an analytic process, depending on the differing sensitivities of a large number of cerebral and retinal systems, each tuned to respond to different components of perception. It was in this atmosphere of hospitality to subsystems and their integration that Zeki discovered specific cells sensitive to wavelength and color in the visual cortex of the monkey, and he found them in much the same area that Verrey had suggested as a color center eighty-five years before. Zeki's discovery seemed to release clinical neurologists from their almost century-long inhibition. Within a few years, scores of new cases of achromatopsia were described, and it was at last legitimized as a valid neurological condition.

That conceptual bias was responsible for the dismissal and "disappearance" of achromatopsia is confirmed by the completely opposite history of central motion blindness, an even rarer condition that was described in a single case by Josef Zihl and his colleagues in 1983.[8] Zihl's patient could see people or cars at rest, but as soon as they began to move, they disappeared from her consciousness, only to reappear, motionless, in another place. This case, Zeki noted, was "immediately accepted by the neurological . . . and the neurobiological world, without a murmur of dissent . . . in contrast to the more turbulent history of achromatopsia." This dramatic difference stemmed from the profound change in intellectual climate which had come about in the years immediately before. In the early 1970s it had been shown that there was a specialized area of motion-sensitive cells in the prestriate cortex of monkeys, and the idea of functional specialization was fully accepted within a decade. There was no longer any conceptual reason for rejecting Zihl's findings—indeed, quite the contrary; they were embraced with delight, as a superb piece of clinical evidence in consonance with the new climate.

That it is crucially important to take note of exceptions—and not forget them or dismiss them as trivial—was brought out in Wolfgang Köhler's first paper, written in 1913, before his pioneer work in gestalt psy-

8. Zihl's case is described in greater detail in the previous chapter, "The River of Consciousness."

chology. In his paper, "On Unnoticed Sensations and Errors of Judgment," Köhler wrote of how premature simplifications and systemizations in science, psychology in particular, could ossify science and prevent its vital growth. "Each science," he wrote, "has a sort of attic into which things are almost automatically pushed that cannot be used at the moment, that do not quite fit. . . . We are constantly putting aside, unused, a wealth of valuable material [that leads to] the blocking of scientific progress."[9]

At the time that Köhler wrote this, visual illusions were seen as "errors of judgment"—trivial, of no relevance to the workings of the mind-brain. But Köhler would soon show that the opposite was the case, that such illusions constituted the clearest evidence that perception does not just passively "process" sensory stimuli but actively creates large configurations or "gestalts" that organize the entire perceptual field. These insights now lie at the heart of our present understanding of the brain as dynamic and constructive. But it was first necessary to seize on an "anomaly," a phenomenon contrary to the accepted frame of reference, and, by according it attention, to enlarge and revolutionize that frame of reference.

9. Darwin remarked on the importance of "negative instances" or "exceptions" and how crucial it is to make immediate note of them, for otherwise they are "sure to be forgotten."

Can we draw any lessons from the examples I have been discussing? I believe we can. One might first invoke the concept of prematurity here and see the nineteenth-century observations of Herschel, Weir Mitchell, Tourette, and Verrey as having come before their times, so that they could not be integrated into contemporary conceptions. Gunther Stent, considering "prematurity" in scientific discovery in 1972, wrote, "A discovery is premature if its implications cannot be connected by a series of simple logical steps to canonical, or generally accepted, knowledge." He discussed this in relation to the classic case of Gregor Mendel, whose work on plant genetics was so far ahead of its time, as well as the lesser-known but fascinating case of Oswald Avery, who discovered DNA in 1944—a discovery totally overlooked because no one could yet appreciate its importance.[10]

Had Stent been a geneticist rather than a molecular biologist, he might have recalled the story of the pioneer

10. Stent's article, "Prematurity and Uniqueness in Scientific Discovery," appeared in *Scientific American* in December 1972. When I visited W. H. Auden in Oxford two months later, he was greatly excited by Stent's article, and we spent much time discussing it. Auden wrote a lengthy reply to Stent, contrasting the intellectual histories of art and science; this was published in the March 1973 *Scientific American.*

geneticist Barbara McClintock, who in the 1940s developed a theory—of so-called jumping genes—which was almost unintelligible to her contemporaries. Thirty years later, when the atmosphere in biology had become more hospitable to such notions, McClintock's insights were belatedly recognized as a fundamental contribution to genetics.

Had Stent been a geologist, he might have given another famous (or infamous) example of prematurity—Alfred Wegener's theory of continental drift, proposed in 1915, forgotten or derided for many years, but then rediscovered forty years later with the rise in plate tectonics theory.

Had Stent been a mathematician, he might even have cited, as an astonishing example of "prematurity," Archimedes's invention of calculus two thousand years before Newton's and Leibniz's.

And had he been an astronomer, he might have spoken not merely of a forgetting but of a most momentous regression in the history of astronomy. Aristarchus, in the third century B.C., clearly established a heliocentric picture of the solar system that was well understood and accepted by the Greeks. (It was further amplified by Archimedes, Hipparchus, and Eratosthenes.) Yet Ptolemy, five centuries later, turned this on its head and proposed a geocentric theory of almost Babylonian complexity. The Ptolemaic darkness, the scotoma, lasted 1,400 years, until a heliocentric theory was reestablished by Copernicus.

Scotoma, surprisingly common in all fields of science, involves more than prematurity; it involves a loss of knowledge, a forgetting of insights that once seemed clearly established, and sometimes a regression to less perceptive explanations. What makes an observation or a new idea acceptable, discussable, memorable? What may prevent it from being so, despite its clear importance and value?

Freud would answer this question by emphasizing resistance: the new idea is deeply threatening or repugnant, and hence is denied full access to the mind. This doubtless is often true, but it reduces everything to psychodynamics and motivation, and even in psychiatry this is not enough.

It is not enough to apprehend something, to "get" something, in a flash. The mind must be able to accommodate it, to retain it. The first barrier lies in allowing oneself to encounter new ideas, to create a mental space, a category with potential connection—and then to bring these ideas into full and stable consciousness, to give them conceptual form, holding them in mind even if they contradict one's existing concepts, beliefs, or categories. This process of accommodation, of spaciousness of mind, is crucial in determining whether an idea or discovery will take hold and bear fruit or whether it will be forgotten, fade, and die without issue.

We have spoken of discoveries or ideas so premature as to be almost without connection or context, hence unintelligible, or ignored, at the time and of other ideas passionately, even ferociously, contested in the necessary but often brutal agon of science. The history of science and medicine has taken much of its shape from intellectual rivalries that force scientists to confront both anomalies and deeply held ideologies. Such competition, in the form of open and straightforward debate and trial, is essential to scientific progress.[11] This is "clean" science, in which friendly or collegial competition encourages an advance in understanding—but there is a good deal of "dirty" science too, in which competition and personal rivalry become malignant and obstructive.

If one aspect of science lies in the realm of competition and rivalry, another one springs from epistemo-

11. Darwin was at pains to say that he had no forerunners, that the idea of evolution was not in the air. Newton, despite his famous comment about "standing on the shoulders of giants," also denied any forerunners. This "anxiety of influence" (which Harold Bloom has discussed powerfully in regard to the history of poetry) is a potent force in the history of science as well. In order to successfully develop and unfold one's own ideas, one may have to believe that others are wrong; one may have to, as Bloom insists, misunderstand others and (perhaps unconsciously) react against them. ("Every talent," Nietzsche writes, "must unfold itself in fighting.")

logical misunderstanding and schism, often of a very fundamental sort. Edward O. Wilson describes in his autobiography, *Naturalist,* how James Watson regarded Wilson's early work in entomology and taxonomy as no more than "stamp collecting." Such a dismissive attitude was almost universal among molecular biologists in the 1960s. (Ecology, similarly, was scarcely allowed status as a "real" science in those days and is still seen as much "softer" than, for example, molecular biology—a mind-set that is only now beginning to shift.)

Darwin often remarked that no man could be a good observer unless he was an active theorizer as well. As Darwin's son Francis wrote, his father seemed "charged with theorising power ready to flow into any channel on the slightest disturbance, so that no fact, however small, could avoid releasing a stream of theory, and thus the fact became magnified into importance." Theory, though, can be a great enemy of honest observation and thought, especially when it hardens into unstated, perhaps unconscious, dogma or assumption.

Undermining one's existing beliefs and theories can be a very painful, even terrifying, process—painful because our mental lives are sustained, consciously or unconsciously, by theories, sometimes invested with the force of ideology or delusion.

In extreme cases scientific debate can threaten to destroy the belief systems of one of the antagonists and with this, perhaps, the beliefs of an entire culture. Darwin's publication of the *Origin* in 1859 instigated furious debates between science and religion (embodied in the conflict between Thomas Huxley and Bishop Wilberforce), and the violent but pathetic rearguard actions of Agassiz, who felt that his lifework, and his sense of a creator, were annihilated by Darwin's theory. The anxiety of obliteration was such that Agassiz actually went to the Galápagos himself and tried to duplicate Darwin's experience and findings in order to repudiate his theory.[12]

Philip Henry Gosse, a great naturalist who was also deeply devout, was so torn by the debate over evolution by natural selection that he was driven to publish an extraordinary book, *Omphalos,* in which he maintained that fossils do not correspond to any creatures that ever lived, but were merely put in the rocks by the Creator to rebuke our curiosity—an argument which had the unusual distinction of infuriating zoologists and theologians in equal measure.

It has sometimes surprised me that chaos theory was not discovered or invented by Newton or Galileo; they must have been perfectly familiar, for example, with

12. Darwin himself was often appalled by the very mechanism of nature whose workings he saw so clearly. He expressed this in a letter which he wrote to his friend Joseph Hooker in 1856: "What a book a Devil's Chaplain might write on the clumsy, wasteful, blundering low and horribly cruel works of nature!"

the phenomena of turbulence and eddies which are constantly seen in daily life (and so consummately portrayed by Leonardo). Perhaps they avoided thinking of their implications, foreseeing these as potential infractions of a rational, lawful, orderly Nature.

This is much what Henri Poincaré felt more than two centuries later, when he became the first to investigate the mathematical consequences of chaos: "These things are so bizarre that I cannot bear to contemplate them." Now we find the patterns of chaos beautiful—a new dimension of nature's beauty—but this was certainly not how it originally seemed to Poincaré.

The most famous example of such repugnance in our own century is, of course, Einstein's violent distaste for the seemingly irrational nature of quantum mechanics. Even though he himself had been one of the very first to demonstrate quantum processes, he refused to consider quantum mechanics anything more than a superficial representation of natural processes, which would give way, with deeper insight, to a more harmonious and orderly one.

⁓⁓⁓⁓

With great scientific advances, there is often both fortuity and inevitability. If Watson and Crick had not cracked the double helix of DNA in 1953, Linus Pauling would almost certainly have done so. The structure of DNA, one might say, was ready to be discovered, though who did it, and how, and exactly when, remained unpredictable.

The greatest creative achievements arise not only from extraordinary, gifted men and women but from their being confronted by problems of enormous universality and magnitude. The sixteenth century was a century of genius not because there were more geniuses around but because the understanding of the laws of the physical world, more or less petrified since the time of Aristotle, was beginning to yield to the insights of Galileo and others who believed that the language of Nature was mathematics. In the seventeenth century, similarly, the time was ripe for the invention of calculus, and it was devised by both Newton and Leibniz almost simultaneously, though in entirely different ways.

In Einstein's time, it was increasingly clear that the old mechanical, Newtonian worldview was insufficient to explain various phenomena—among them the photoelectric effect, Brownian motion, and the change of mechanics near the speed of light—and had to collapse and leave a rather frightening intellectual vacuum before a radically new concept could be born.

But Einstein also took pains to say that a new theory does not invalidate or supersede the old but rather "allows us to regain our old concepts from a higher level." He expanded this notion in a famous simile:

> To use a comparison, we could say that creating a new theory is not like destroying an old barn and erecting a skyscraper in its place. It is rather like climbing a mountain, gaining new and wider views, discovering unexpected connections between our starting point and its rich environment. But the point from which we started out still exists and can be seen, although it appears smaller and forms a tiny part of our broad view gained by the mastery of the obstacles on our adventurous way up.

Helmholtz, in his memoir *On Thought in Medicine,* also used the image of a mountain climb (he was an ardent alpinist), describing the climb as anything but linear. One cannot see in advance, he wrote, how to climb a mountain; it can only be climbed by trial and error. The intellectual mountaineer makes false starts, turns in to blind alleys, finds himself in untenable positions and often has to backtrack, descend, and start again. Slowly and painfully, with innumerable errors and corrections, he makes his zigzag way up the mountain. It is only when he reaches the summit that he will see that there was, in fact, a direct route, a "royal road," to the top. In presenting his ideas, Helmholtz says, he takes his readers along this royal road, but it bears no resemblance to the crooked and tortuous processes by which he constructed a path for himself.

Often there is some intuitive and inchoate vision of what must be done, and this vision, once glimpsed, drives the intellect forward. Thus Einstein at the age of fifteen had fantasies about riding a light beam and ten years later developed the theory of special relativity, going from a boy's dream to the grandest of theories. Was the achievement of the theory of special relativity, and then of general relativity, part of an ongoing, inevitable historical process? Or the result of a singularity, the advent of a unique genius? Would relativity have been conceived in Einstein's absence? And how quickly would relativity have been accepted had it not been for the solar eclipse of 1917, which, by a rare chance, allowed the theory to be confirmed by accurate observation of the effect of the sun's gravity on light? One senses the fortuitous here—and, not trivially, a requisite level of technology, one which could measure Mercury's orbit accurately. Neither "historical process" nor "genius" is an adequate explanation—each glosses over the complexity, the chancy nature, of reality.

"Chance favors the prepared mind," as Claude Bernard famously wrote, and Einstein was, of course, intensely alert, primed to perceive and seize whatever he could use. But if Riemann and other mathematicians had not developed non-Euclidean geometries (they had been worked out as pure abstract constructions, with no notion that they might be appropriate to any physical model of the world), Einstein would not have had the intellectual techniques available to move from a vague vision to a fully developed theory.

A number of isolated, autonomous, individual factors must converge before the seemingly magical act of a creative advance, and the absence (or insufficient development) of any one may suffice to prevent it. Some of these factors are worldly ones—sufficient funding and opportunity, health and social support, the era into which one was born. Others have to do with innate personality and intellectual strengths or weaknesses.

In the nineteenth century, an era of naturalistic description and phenomenological passion for detail, a concrete habit of mind seemed highly appropriate, and an abstract or ratiocinating one was suspect—an attitude beautifully brought out by William James in his famous essay on Louis Agassiz, the eminent biologist and natural historian:

> The only man he really loved and had use for was the man who could bring him facts. To see facts, not to argue or [reason], was what life meant for him; and I think he often positively loathed the ratiocinating type of mind. . . . The extreme rigor of his devotion to this concrete method of learning was the natural consequence of his own peculiar type of intellect, in which the capacity for abstraction and causal reasoning and tracing chains of consequences from

> hypotheses was so much less developed than the genius for acquaintance with vast volumes of detail, and for seizing upon analogies and relations of the more proximate and concrete kind.

James describes how the young Agassiz, coming to Harvard in the mid-1840s, "studied the geology and fauna of a continent, trained a generation of zoologists, founded one of the chief museums of the world, gave a new impulse to scientific education in America"—and all this through his passionate love of phenomena and facts, of fossils and living forms, his lyrical concreteness of mind, his scientific and religious sense of a divine system, a whole. But then there came a transformation: zoology itself was changing from a natural history, intent on wholes—species and forms and their taxonomic relationships—to studies in physiology, histology, chemistry, pharmacology, a new science of the micro, of mechanisms and parts abstracted from a sense of the organism and its organization as a whole. Nothing was more exciting, more potent than this new science, and yet it was clear that something was being lost, too. It was a transformation to which Agassiz's mind could not well adapt, and he was pushed, in his later years, away from the center of scientific thought, becoming an eccentric and tragic figure.[13]

13. Humphry Davy, like Agassiz, was a genius of concreteness and analogical thinking. He lacked the power of abstract generalization that was so strong in his contemporary John Dalton (it is to Dalton that we owe the foundations of atomic theory) and the massive sys-

The huge role of contingency, of sheer luck (good or bad), it seems to me, is even more evident in medicine than in science, for medicine often depends crucially on rare and unusual, perhaps unique, cases being encountered by the right person at the right time.

Cases of prodigious memory are naturally rare, and the Russian Shereshevsky was among the most remarkable of these. But he would be remembered now (if at all) as merely "another case of prodigious memory" had it not been for the chance of meeting A. R. Luria, himself a prodigy of clinical observation and insight. It required the genius of a Luria, and his thirty-year-long exploration of Shereshevsky's mental processes, to produce the unique insights of Luria's great book *The Mind of a Mnemonist.*

Hysteria, by contrast, is not uncommon, and has been well described since the eighteenth century. But it was

---

tematic powers of his contemporary Berzelius. Davy hence descended from his idolized position as "the Newton of chemistry" in 1810 to being almost marginal fifteen years later. The rise of organic chemistry, with Wöhler's synthesis of urea in 1828—a new realm in which Davy had no interest or understanding—immediately started to displace the "old" inorganic chemistry and added to Davy's sense of being outmoded in his last years.

Jean Améry, in his powerful book *On Aging,* speaks of how tormenting the sense of irrelevance or obsolescence may be, in particular the sense of being *intellectually* outmoded through the rise of new methods, theories, or systems. Such outmoding in science can occur almost instantly when there is a major shift of thought.

not plumbed psychodynamically until a brilliant, articulate hysteric encountered the original genius of the young Freud and his friend Breuer. Would psychoanalysis, one wonders, ever have got going without Anna O. meeting up with the specially receptive, prepared minds of Freud and Breuer? (I am sure that it would have, but later, and in a different way.)

Could the history of science—like life—be rerun quite differently? Does the evolution of ideas resemble the evolution of life? Assuredly we see sudden explosions of activity, when enormous advances are made in a very short time. This was so for molecular biology in the 1950s and 1960s and for quantum physics in the 1920s, and a similar burst of fundamental work has occurred in neuroscience over the past few decades. Sudden bursts of discovery change the face of science, and these are often followed by long periods of consolidation and relative stasis. I am reminded of the picture of "punctuated equilibrium" given us by Niles Eldredge and Stephen Jay Gould and wonder if there is at least an analogy here to a natural evolutionary process.

Ideas, like living creatures, may arise and flourish, going in all directions, or abort and become extinct, in completely unpredictable ways. Gould was fond of saying that if the evolution of life on earth could be replayed, it would be wholly different the second time around. Suppose that John Mayow had indeed discovered oxygen in the 1670s or that Babbage's theoretical Difference

Engine—a computer—had been built when he proposed it in 1822; might the course of science have been quite different? This is the stuff of fantasy, of course, but fantasy that brings home a sense that science is not an ineluctable process but contingent in the extreme.

Bibliography

Améry, Jean. 1994. *On Aging.* Bloomington: Indiana University Press.

Arendt, Hannah. 1971. *The Life of the Mind.* New York: Harcourt.

Armitage, F. P. 1906. *A History of Chemistry.* London: Longmans Green.

Bartlett, Frederic C. 1932. *Remembering: A Study in Experimental and Social Psychology.* Cambridge, U.K.: Cambridge University Press.

Bergson, Henri. 1911. *Creative Evolution.* New York: Henry Holt.

Bernard, Claude. 1865. *An Introduction to the Study of Experimental Medicine.* London: Macmillan.

Bleuler, Eugen. 1911/1950. *Dementia Praecox; or, The Group of Schizophrenias.* Oxford: International Universities Press.

Bloom, Harold. 1973. *The Anxiety of Influence.* Oxford: Oxford University Press.

Braun, Marta. 1992. *Picturing Time: The Work of Etienne-Jules Marey (1830–1904).* Chicago: University of Chicago Press.

Brock, William H. 1993. *The Norton History of Chemistry.* New York: W. W. Norton.

Browne, Janet. 2002. *Charles Darwin: The Power of Place.* New York: Alfred A. Knopf.

Chamovitz, Daniel. 2012. *What a Plant Knows: A Field Guide to the Senses.* New York: Scientific American/Farrar, Straus and Giroux.

Changeux, Jean-Pierre. 2004. *The Physiology of Truth: Neuroscience and Human Knowledge.* Cambridge, Mass.: Harvard University Press.

Coleridge, Samuel Taylor. 1817. *Biographia Literaria.* London: Rest Fenner.

Crick, Francis. 1994. *The Astonishing Hypothesis: The Scientific Search for the Soul.* New York: Charles Scribner.

Damasio, Antonio. 1999. *The Feeling of What Happens: Body and Emotion in the Making of Consciousness.* New York: Harcourt Brace.

Damasio, A., T. Yamada, H. Damasio, J. Corbett, and J. McKee. 1980. "Central Achromatopsia: Behavioral, Anatomic, and Physiologic Aspects." *Neurology* 30 (10): 1064–71.

Damasio, Antonio, and Gil B. Carvalho. 2013. "The Nature of Feelings: Evolutionary and Neurobiological Origins." *Nature Reviews Neuroscience* 14, February.

Darwin, Charles. 1859. *On the Origin of Species by Means of Natural Selection; or, The Preservation of Favoured Races in the Struggle for Life.* London: John Murray.

———. 1862. *On the Various Contrivances by Which British and Foreign Orchids Are Fertilised by Insects.* London: John Murray.

———. 1871. *The Descent of Man, and Selection in Relation to Sex.* London: John Murray.

———. 1875. *On the Movements and Habits of Climbing Plants.* London: John Murray. Linnaean Society paper, originally published in 1865.

———. 1875. *Insectivorous Plants.* London: John Murray.

———. 1876. *The Effects of Cross and Self Fertilisation in the Vegetable Kingdom.* London: John Murray.

———. 1877. *The Different Forms of Flowers on Plants of the Same Species.* London: John Murray.

———. 1880. *The Power of Movement in Plants.* London: John Murray.

———. 1881. *The Formation of Vegetable Mould, Through the Action of Worms, with Observations on Their Habits.* London: John Murray.

Darwin, Erasmus. 1791. *The Botanic Garden: The Loves of the Plants.* London: J. Johnson.

Darwin, Francis, ed. 1887. *The Autobiography of Charles Darwin.* London: John Murray.

Dobzhansky, Theodosius. 1973. "Nothing in Biology Makes Sense Except in the Light of Evolution." *American Biology Teacher* 35 (3): 125–29.

Donald, Merlin. 1993. *Origins of the Modern Mind.* Cambridge, Mass.: Harvard University Press.

Doyle, Arthur Conan. 1887. *A Study in Scarlet.* London: Ward, Lock.

———. 1892. *The Adventures of Sherlock Holmes.* London: George Newnes.

———. 1893. "The Adventure of the Final Problem." In *The Memoirs of Sherlock Holmes.* London: George Newnes.

———. 1905. *The Return of Sherlock Holmes.* London: George Newnes.

Edelman, Gerald M. 1987. *Neural Darwinism: The Theory of Neuronal Group Selection.* New York: Basic Books.

———. 1989. *The Remembered Present: A Biological Theory of Consciousness.* New York: Basic Books.

———. 2004. *Wider Than the Sky: The Phenomenal Gift of Consciousness.* New York: Basic Books.

Efron, Daniel H., ed. 1970. *Psychotomimetic Drugs: Proceedings of a Workshop . . . Held at the University of California, Irvine, on January 25–26, 1969.* New York: Raven Press.

Einstein, Albert, and Leopold Infeld. 1938. *The Evolution of Physics.* Cambridge, U.K.: Cambridge University Press.

Flannery, Tim. 2013. "They're Taking Over!" *New York Review of Books,* Sept. 26.

Freud, Sigmund. 1891/1953. *On Aphasia: A Critical Study.* Oxford: International Universities Press.

———. 1901/1990. *The Psychopathology of Everyday Life.* New York: W. W. Norton.

Freud, Sigmund, and Josef Breuer. 1895/1991. *Studies on Hysteria.* New York: Penguin.

Friel, Brian. 1994. *Molly Sweeney.* New York: Plume.

Gooddy, William. 1988. *Time and the Nervous System.* New York: Praeger.

Gosse, Philip Henry. 1857. *Omphalos: An Attempt to Untie the Geological Knot.* London: John van Voorst.

Gould, Stephen Jay. 1990. *Wonderful Life.* New York: W. W. Norton.

Greenspan, Ralph J. 2007. *An Introduction to Nervous Systems.* Cold Spring Harbor, N.Y.: Cold Spring Harbor Laboratory Press.

Hadamard, Jacques. 1945. *The Psychology of Invention in the Mathematical Field.* Princeton, N.J.: Princeton University Press.

Hales, Stephen. 1727. *Vegetable Staticks.* London: W. and J. Innys.

Hanlon, Roger T., and John B. Messenger. 1998. *Cephalopod Behaviour.* Cambridge, U.K.: Cambridge University Press.

Hebb, Donald. 1949. *The Organization of Behavior: A Neuropsychological Theory.* New York: Wiley.

Helmholtz, Hermann von. 1860/1962. *Treatise on Physiological Optics.* New York: Dover.

———. 1877/1938. *On Thought in Medicine.* Baltimore: Johns Hopkins Press.

Herrmann, Dorothy. 1998. *Helen Keller: A Life.* Chicago: University of Chicago Press.

Herschel, J. F. W. 1858/1866. "On Sensorial Vision." In *Familiar Lectures on Scientific Subjects.* London: Alexander Strahan.

Holmes, Richard. 1989. *Coleridge: Early Visions, 1772–1804.* New York: Pantheon.

———. 2000. *Coleridge: Darker Reflections, 1804–1834.* New York: Pantheon.

Jackson, John Hughlings. 1932. *Selected Writings.* Vol. 2. Edited by James Taylor, Gordon Holmes, and F. M. R. Walshe. London: Hodder and Stoughton.

James, William. 1890. *The Principles of Psychology.* London: Macmillan.

———. 1896/1984. *William James on Exceptional Mental States: The 1896 Lowell Lectures.* Edited by Eugene Taylor. Amherst: University of Massachusetts Press.

———. 1897. *Louis Agassiz: Words Spoken by Professor William James at the Reception of the American Society of Naturalists by the President and Fellows of Harvard College, at Cambridge, on December 30, 1896.* Cambridge, Mass.: printed for the university.

Jennings, Herbert Spencer. 1906. *Behavior of the Lower Organisms.* New York: Columbia University Press.

Kandel, Eric R. 2007. *In Search of Memory: The Emergence of a New Science of Mind.* New York: W. W. Norton.

Keynes, John Maynard. 1946. "Newton, the Man." *http://www-history.mcs.st-and.ac.uk/Extras/Keynes_Newton.html.*

Knight, David. 1992. *Humphry Davy: Science and Power.* Cambridge, U.K.: Cambridge University Press.

Koch, Christof. 2004. *The Quest for Consciousness: A Neurobiological Approach.* Englewood, Colo.: Roberts.

Köhler, Wolfgang. 1913/1971. "On Unnoticed Sensations and Errors of Judgment." In *The Selected Papers of Wolfgang Köhler,* edited by Mary Henle. New York: Liveright.

Kohn, David. 2008. *Darwin's Garden: An Evolutionary Adventure.* New York: New York Botanical Garden.

Kraepelin, Emil. 1904. *Lectures on Clinical Psychiatry.* New York: William Wood.

Lappin, Elena. 1999. "The Man with Two Heads." *Granta* 66:7–65.

Leont'ev, A. N., and A. V. Zaporozhets. 1960. *Rehabilitation of Hand Function.* Oxford: Pergamon Press.

Libet, Benjamin, C. A. Gleason, E. W. Wright, and D. K. Pearl. 1983. "Time of Conscious Intention to Act in Relation to Onset of Cerebral Activity (Readiness-Potential): The Unconscious Initiation of a Freely Voluntary Act." *Brain* 106:623–42.

Liveing, Edward. 1873. *On Megrim, Sick-Headache, and Some Allied Disorders: A Contribution to the Pathology of Nerve-Storms.* London: Churchill.

Loftus, Elizabeth. 1996. *Eyewitness Testimony.* Cambridge, Mass.: Harvard University Press.

Lorenz, Konrad. 1981. *The Foundations of Ethology.* New York: Springer.

Luria, A. R. 1968. *The Mind of a Mnemonist.* Reprint, Cambridge, Mass.: Harvard University Press.

———. 1973. *The Working Brain: An Introduction to Neuropsychology.* New York: Basic Books.

———. 1979. *The Making of Mind.* Cambridge, Mass.: Harvard University Press.

Meige, Henri, and E. Feindel. 1902. *Les tics et leur traitement.* Paris: Masson.

Meynert, Theodor. 1884/1885. *Psychiatry: A Clinical Treatise on Diseases of the Fore-brain.* New York: G. P. Putnam's Sons.

Michaux, Henri. 1974. *The Major Ordeals of the Mind and the Countless Minor Ones.* London: Secker and Warburg.

Mitchell, Silas Weir. 1872/1965. *Injuries of Nerves and Their Consequences.* New York: Dover.

Mitchell, Silas Weir, W. W. Keen, and G. R. Morehouse. 1864. *Reflex Paralysis.* Washington, D.C.: Surgeon General's Office.

Modell, Arnold. 1993. *The Private Self.* Cambridge, Mass.: Harvard University Press.

Moreau, Jacques-Joseph. 1845/1973. *Hashish and Mental Illness.* New York: Raven Press.

Nietzsche, Friedrich. 1882/1974. *The Gay Science.* Translated by Walter Kaufmann. New York: Vintage Books.

Noyes, Russell, Jr., and Roy Kletti. 1976. "Depersonalization in the Face of Life-Threatening Danger: A Description." *Psychiatry* 39 (1): 19–27.

Orwell, George. 1949. *Nineteen Eighty-Four.* London: Secker and Warburg.

Pinter, Harold. 1994. *Other Places: Three Plays.* New York: Grove Press.

Pribram, Karl H., and Merton M. McGill. 1976. *Freud's "Project" Re-assessed.* New York: Basic Books.

Romanes, George John. 1883. *Mental Evolution in Animals.* London: Kegan Paul, Trench.

———. 1885. *Jelly-Fish, Star-Fish, and Sea-Urchins: Being a*

Research on Primitive Nervous Systems. London: Kegan Paul, Trench.
Sacks, Oliver. 1973. *Awakenings.* New York: Doubleday.
———. 1984. *A Leg to Stand On.* New York: Summit Books.
———. 1985. *The Man Who Mistook His Wife for a Hat.* New York: Summit Books.
———. 1992. *Migraine.* Rev. ed. New York: Vintage Books.
———. 1993. "Humphry Davy: The Poet of Chemistry." *New York Review of Books,* Nov. 4.
———. 1993. "Remembering South Kensington." *Discover* 14(11): 78-80.
———. 1995. *An Anthropologist on Mars.* New York: Alfred A. Knopf.
———. 1996. *The Island of the Colorblind.* New York: Alfred A. Knopf.
———. 2001. *Uncle Tungsten.* New York: Alfred A. Knopf.
———. 2007. *Musicophilia: Tales of Music and the Brain.* New York: Alfred A. Knopf.
———. 2012. *Hallucinations.* New York: Alfred A. Knopf.
Sacks, O. W., O. Fookson, M. Berkinblit, B. Smetanin, R. M. Siegel, and H. Poizner. 1993. "Movement Perturbations due to Tics Do Not Affect Accuracy on Pointing to Remembered Locations in 3-D Space in a Subject with Tourette's Syndrome." *Society for Neuroscience Abstracts* 19 (1): item 228.7.
Schacter, Daniel L. 1996. *Searching for Memory: The Brain, the Mind, and the Past.* New York: Basic Books.
———. 2001. *The Seven Sins of Memory.* New York: Houghton Mifflin.
Shenk, David. 2001. *The Forgetting: Alzheimer's: Portrait of an Epidemic.* New York: Doubleday.

Sherrington, Charles. 1942. *Man on His Nature.* Cambridge, U.K.: Cambridge University Press.

Solnit, Rebecca. 2003. *River of Shadows: Eadweard Muybridge and the Technological Wild West.* New York: Viking.

Spence, Donald P. 1982. *Narrative Truth and Historical Truth: Meaning and Interpretation in Psychoanalysis.* New York: Norton.

Sprengel, Christian Konrad. 1793/1975. *The Secret of Nature in the Form and Fertilization of Flowers Discovered.* Washington, D.C.: Saad.

Stent, Gunther. 1972. "Prematurity and Uniqueness in Scientific Discovery." *Scientific American* 227 (6): 84–93.

Tourette, Georges Gilles de la. 1885. "Étude sur une affection nerveuse caractérisée par de l'incoordination motrice accompagnée d'écholalie et de copralalie." *Archives de Neurologie* (Paris) 9.

Twain, Mark. 1917. *Mark Twain's Letters,* vol. 1. Ed. Albert Bigelowe Paine. New York: Harper & Bros.

———. 2006. *Mark Twain Speaking.* Town City: University of Iowa Press.

Vaughan, Ivan. 1986. *Ivan: Living with Parkinson's Disease.* London: Macmillan.

Verrey, Louis. 1888. "Hémiachromatopsie droite absolue." *Archives d'Ophthamologie* (Paris) 8: 289–300.

Wade, Nicholas J. 2000. *A Natural History of Vision.* Cambridge, Mass.: MIT Press.

Weinstein, Arnold. 2004. *A Scream Goes Through the House: What Literature Teaches Us About Life.* New York: Random House.

Wells, H. G. 1927. *The Short Stories of H. G. Wells.* London: Ernest Benn.

Wiener, Norbert. 1953. *Ex-Prodigy: My Childhood and Youth.* New York: Simon & Schuster.

Wilkomirski, Binjamin. 1996. *Fragments: Memories of a Wartime Childhood.* New York: Schocken.

Wilson, Edward O. 1994. *Naturalist.* Washington, D.C.: Island Press.

Zeki, Semir. 1990. "A Century of Cerebral Achromatopsia." *Brain* 113:1721–77.

Zihl, J., D. von Cramon, and N. Mai. 1983. "Selective Disturbance of Movement Vision after Bilateral Brain Damage." *Brain* 106 (2): 313–40.

Index

A Note About the Author

Oliver Sacks was born in London in 1933. He studied medicine at Oxford, followed by a residency at UCLA. For the next fifty years, he worked as a neurologist at various institutions in New York City for the chronically ill, including Beth Abraham Hospital in the Bronx and several nursing homes run by the Little Sisters of the Poor.

The New York Times referred to Sacks as "the poet laureate of medicine." He is best known for his collections of neurological case histories, including *The Man Who Mistook His Wife for a Hat, Musicophilia, An Anthropologist on Mars,* and *Hallucinations.* "Again and again," said the *Los Angeles Times,* "Sacks invites readers to imagine their way into minds unlike their own, encouraging a radical form of empathy."

Awakenings, his 1973 book about a group of patients who had survived the great encephalitis lethargica epidemic of the early twentieth century, inspired the 1990 Academy Award–nominated feature film starring Robert De Niro and Robin Williams.

Dr. Sacks was a frequent contributor to *The New Yorker* and *The New York Review of Books* and many other journals. He was a member of the Royal College of Physicians, the American Academy of Arts and Letters, and the American Academy of Arts and Sciences. In 2008, Queen Elizabeth II named him a Commander of the British Empire.

Dr. Sacks served as a board member of the New York Botanical Garden, which awarded him their Gold Medal in 2011.

The asteroid 84928 Oliversacks was named in honor of his seventy-fifth birthday in 2008.

Dr. Sacks died in New York City in 2015, a few months after the publication of his memoir, *On the Move*.

For more information about Dr. Sacks and the Oliver Sacks Foundation, please visit *www.oliversacks.com*.

A Note on the Type

The text of this book was composed in Trump Mediaeval. Designed by Professor Georg Trump (1896–1985) in the mid-1950s, Trump Mediaeval was cut and cast by the C. E. Weber Type Foundry of Stuttgart, Germany. The roman letterforms are based on classical prototypes, but Professor Trump has imbued them with his own unmistakable style. The italic letterforms, unlike those of so many other typefaces, are closely related to their roman counterparts. The result is a truly contemporary type, notable for both its legibility and its versatility.

Typeset by Scribe, Philadelphia, Pennsylvania

Printed and bound by LSC Communications,
Harrisonburg, Virginia

Designed by Iris Weinstein